Think Like a Rocket Scientist

Think Like a Rocket Scientist

Simple Strategies You Can Use to Make Giant Leaps in Work and Life

OZAN VAROL

PUBLICAFFAIRS

New York

PublicAffairs
Hachette Book Group
1290 Avenue of the Americas, New York, NY 10104
www.publicaffairsbooks.com
@Public_Affairs

Printed in the United States of America

First Edition: April 2020

Published by PublicAffairs, an imprint of Perseus Books, LLC, a subsidiary of Hachette Book Group, Inc. The PublicAffairs name and logo is a trademark of the Hachette Book Group.

The Hachette Speakers Bureau provides a wide range of authors for speaking events. To find out more, go to www.hachettespeakersbureau.com or call (866) 376-6591.

The publisher is not responsible for websites (or their content) that are not owned by the publisher.

Print book interior design by Trish Wilkinson.

Library of Congress Cataloging-in-Publication Data

Names: Varol, Ozan O., 1981– author.
Title: Think like a rocket scientist : simple strategies you can use to make giant leaps in work and life / Ozan Varol.
Identifiers: LCCN 2019041750 | ISBN 9781541762596 (hardcover) | ISBN 9781541762619 (ebook)
Subjects: LCSH: Creative ability in business. | Creative ability. | Scientific ability.
Classification: LCC HD53 .V37 2020 | DDC 650.1—dc23
LC record available at https://lccn.loc.gov/2019041750

ISBNs: 978-1-5417-6259-6 (hardcover), 978-1-5417-6261-9 (ebook), 978-1-5417-5710-3 (international)

LSC-C

10 9 8 7 6 5 4 3 2 1

To Kathy, my cosmic constant

CONTENTS

INTRODUCTION

IN SEPTEMBER 1962, President John F. Kennedy stood before a packed Rice University stadium and pledged to land a man on the Moon and return him safely to the Earth before the decade was out. It was an incredibly ambitious promise—the original moonshot.

When Kennedy gave his speech, numerous technological requirements for a Moon landing hadn't even been developed. No American astronauts had worked outside a spacecraft.[1] Two spacecraft had never docked together in space.[2] The National Aeronautics and Space Administration (NASA) didn't know whether the lunar surface was sufficiently solid to support a lander or whether the communications systems would work on the Moon.[3] In the words of one NASA executive, we didn't even know "how to do [Earth] orbit determination, much less project orbits to the Moon."[4]

Getting into orbit around the Moon—not to mention landing on it—required mind-blowing precision. It was like throwing a dart at a peach twenty-eight feet away and scraping the fuzz without touching the body.[5] What's more, the peach—the Moon—would be in rapid motion, darting through space. On reentry to the Earth, the spacecraft would have to enter the atmosphere at just the right angle—tantamount to locating one particular ridge on a coin of 180 ridges—to avoid grinding too hard against the atmosphere and burning to a crisp or skidding across it like a stone skipping on water.[6]

For a politician, Kennedy was surprisingly candid about the challenges ahead. The giant rocket to take the astronauts to the Moon, he explained, would be "made of new metal alloys, some of which have not yet been invented, capable of standing heat and stresses several times more than have ever been experienced, fitted together with a precision better than the finest watch" and sent "on an untried mission, to an unknown celestial body."[7]

Yes, even the metals needed to build the rocket hadn't been invented.

We jumped into the cosmic void and hoped we would grow wings on the way up.

Miraculously, the wings sprouted. In 1969, less than seven years after Kennedy's pledge, Neil Armstrong took his giant leap for mankind. A child who was six years old when the Wright brothers took their first powered flight—lasting all of twelve seconds and moving 120 feet—would have been seventy-two when flight became powerful enough to put a man on the Moon and return him safely to the Earth.

This giant leap—taken within a human lifespan—is often hailed as the triumph of technology. But it's not. Rather, it's the great triumph of a certain thought process rocket scientists used to turn the impossible into the possible. It's the same thought process that has allowed these scientists to score dozens of interplanetary holes in one with supersonic spacecraft, sending them millions of miles through outer space and landing them on a precise spot. It's the same thought process that brings humanity closer and closer to colonizing other planets and becoming an interplanetary species. And it's the same thought process that will make affordable commercial space tourism the new norm.

To think like a rocket scientist is to look at the world through a different lens. Rocket scientists imagine the unimaginable and solve the unsolvable. They transform failures into triumphs and constraints into advantages. They view mishaps as solvable puzzles rather than insurmountable roadblocks. They're moved not by blind conviction but by self-doubt; their goal is not short-term results but long-term breakthroughs. They know that the rules aren't set in stone, the default can be altered, and a new path can be forged.

Some of the insights I'll share in this book are common to all sciences. But the ideas assume a grander scale in rocket science given the stakes involved. With every launch, hundreds of millions of dollars—and for human spaceflight, numerous lives—are at stake.

At its core, a rocket launch is the controlled explosion of a small nuclear bomb—*controlled* being the operative word. A rocket burns with unbelievable fury. One wrong step, one miscalculation, and you can expect the worst. "There are a thousand things that can happen when you go to light a rocket engine," explains SpaceX propulsion chief Tom Mueller, "and only one of them is good."[8]

Everything we take for granted on Earth is turned on its head in space, literally and metaphorically. There are countless points of potential failure in sending delicate spacecraft—made up of millions of parts and hundreds of miles of wiring—barreling through the unforgiving environment of space.[9] When something breaks, as it inevitably does, rocket scientists must isolate the signal from the noise and home in on the potential culprits, which may be in the thousands. What's worse, these problems often occur when the spacecraft is beyond human reach. You can't just pop the hood and have a look inside.

In the modern era, rocket-science thinking is a necessity. The world is evolving at dizzying speed, and we must continuously evolve with it to keep pace. Although not everyone aspires to calculate burn-rate coefficients or orbital trajectories, we all encounter complex and unfamiliar problems in our daily lives. Those who can tackle these problems—without clear guidelines and with the clock ticking—enjoy an extraordinary advantage.

Despite its tremendous benefits, we often assume that thinking like a rocket scientist is beyond the ability of mere mortals without a special kind of genius (hence the common saying "It's not rocket science"). We identify with Elton John's Rocket Man, who, despite being selected for a Mars mission, laments about "all this science I don't understand."[10] We also empathize with Chaim (Charles) Weizmann, the first president of Israel, who once crossed the Atlantic with Albert Einstein. Every morning, they sat for two hours on the ship's deck, as Einstein explained the theory of relativity to him. At the end of the trip, Weizmann said he was "convinced that Einstein understood relativity."[11]

This book won't teach you relativity or the intricate details of rocket propulsion—in other words, the science behind rocket science. You won't find any graphs on these pages. No aptitude for crunching numbers is required. Lurking behind the elusive subject of rocket science are life-changing insights on creativity and critical thinking that anyone can acquire without a PhD in astrophysics. Science, as Carl Sagan put it, is "a way of thinking much more than it is a body of knowledge."[12]

You won't be a rocket scientist by the end of this book. But you'll know how to think like one.

....

THE TERM *ROCKET SCIENCE* is popular jargon. There's no college major called rocket science, no job with the official title Rocket Scientist. Instead, the term is used colloquially to refer to the science and engineering behind space travel, and that's the broad definition I'll use in this book. I'll explore the work of both scientists—the idealistic explorers engaged in research about the cosmos—and engineers, who are the pragmatic designers of hardware that makes spaceflight possible.

I was once one of them. I worked on the operations team for the Mars Exploration Rovers project, which sent two rovers to the red planet in 2003. I planned operations scenarios, helped select landing sites, and wrote code for snapping photos of Mars. To this day, my rocket-science past remains the most interesting part of my résumé. During speaking engagements, the person introducing me inevitably says, "The most intriguing thing about Ozan is that he used to be a rocket scientist." This produces a collective gasp, and the audience promptly forgets whatever it is I'm there to talk about. I can tell what many of them are thinking: *Talk to us about rocket science instead.*

Let's be honest: We have a love affair with rocket scientists. We despise politicians, we mock lawyers, but we adore those lab-coated brainiacs who design rockets and launch them into the cosmic ocean in a perfectly coordinated symphony. Every Thursday evening, *The Big Bang Theory*—a TV show about a group of eccentric astrophysicists—regularly topped the American charts. Tens of millions break out in laughter as Leslie dumps Leonard because he prefers string theory over loop quantum gravity. For three months, more than three million Americans picked *Cosmos* over *The Bachelor* each Sunday night, choosing dark matter and black holes over the drama of a rose ceremony.[13] Movies about rocket science—from *Apollo 13* to *The Martian*, from *Interstellar* to *Hidden Figures*—consistently top box-office charts and collect countless golden statues.

Although we glamorize rocket scientists, there's an enormous mismatch between what they have figured out and what the rest of the world does. Critical thinking and creativity don't come naturally to us. We're hesitant to think big, reluctant to dance with uncertainty, and afraid of failure. These were necessary during the Paleolithic Period, keeping us safe from poisonous foods and predators. But here in the information age, they're bugs.

Companies fail because they stare at the rearview mirror and keep calling the same plays from the same playbook. Instead of risking failure, they stick with the status quo. In our daily lives, we fail to exercise our critical-thinking muscles and instead leave it to others to draw conclusions. As a result, these muscles atrophy over time. Without an informed public willing to question confident claims, democracy decays and misinformation spreads. Once alternative facts are reported and retweeted, they become the truth. Pseudoscience becomes indistinguishable from real science.

With this book, I aim to create an army of non–rocket scientists who approach everyday problems as a rocket scientist would. You'll take ownership of your life. You'll question assumptions, stereotypes, and established patterns of thinking. Where others see roadblocks, you'll see opportunities to bend reality to your will. You'll approach problems rationally and generate innovative solutions that redefine the status quo. You'll come equipped with a tool kit that enables you to spot misinformation and pseudoscience.

You'll forge new paths and figure out ways to overcome the problems of our future.

As business leaders, you'll ask the right questions and use the right set of tools to make decisions. You won't chase trends, adopt the newest fad, or do things simply because your competitors are doing them. You'll explore the edges and accomplish what others thought was impossible. You'll join the ranks of an elite group of institutions that are beginning to adopt rocket-science thinking into their business model. Wall Street now employs so-called financial rocket scientists to turn investing from an art into a science.[14] Rocket-science thinking is also used by leading retailers to help them pick the next hot product in the face of an uncertain marketplace.[15]

This book is relentlessly practical. It doesn't just preach the benefits of thinking like a rocket scientist. It gives you concrete, actionable strategies for putting that thinking to use, whether you're on the launch pad, in the boardroom, or in your living room. To illustrate how broadly these principles apply, the book weaves gripping anecdotes from rocket science with comparable episodes from history, business, politics, and law to illustrate the rocket-science mindset.

To help you put these principles into action, I've created several free resources on my website, which is an important extension of this book. Visit ozanvarol.com/rocket to find the following:

- A summary of key points from each chapter
- Worksheets, challenges, and other exercises to help you implement the strategies discussed in the book
- A place to sign up for my weekly newsletter, where I share additional tips and resources that reinforce the principles in the book (readers call it "the one email I look forward to each week")
- My personal email address so you can share comments or just say hello

....

ALTHOUGH MY NAME appears on the front cover, this book stands on many shoulders. It draws on my experience working on the operations team for the Mars Exploration Rovers mission, interviews I conducted with numerous rocket scientists, and decades of research in diverse fields, including science and business. I travel frequently to speak about rocket-science thinking to professionals across many industries—law, retail, pharmaceuticals, financial services, to name a few—continually refining my own thinking about how these principles apply in other fields.

I chose to feature nine main principles from rocket science in this book. I left other ideas on the cutting-room floor, focusing on those that have the most relevance beyond space exploration. I'll explain where scientists meet these ideals and where they fall short. You'll learn from the triumphs and tribulations of rocket science—its proudest moments as well as its catastrophes.

Just like rockets, this book comes in stages. The first stage—launch—is dedicated to igniting your thinking. Breakthrough thinking is fraught with uncertainty, so we'll start there. I'll share with you strategies that rocket scientists use to dance with uncertainty and convert it to their advantage. I'll then turn to reasoning from first principles—the ingredient behind every revolutionary innovation. You'll discover the biggest mistake businesses make in generating ideas; how invisible rules constrain your thinking; and why subtracting, rather than adding, is the key to originality. We'll then cover thought experiments and moonshot thinking—strategies used by rocket scientists, innovative businesses, and world-class performers to transform themselves from passive observers to active interveners in their reality. Along the way, you'll learn why it's safer to fly closer to the Sun, how the use of a single word can boost creativity, and what you should do first in tackling an audacious goal.

The second stage—accelerate—is focused on propelling the ideas you created in the first stage. We'll first explore how to reframe and refine your ideas and why finding the right answer begins with asking the right question. We'll then examine how to spot the flaws in your ideas by switching your default from convincing others you're right to proving yourself wrong.

I'll reveal how to test and experiment like a rocket scientist to make sure your ideas have the best shot at landing. Along the way, you'll discover an unstoppable astronaut-training strategy that you can use to nail your next presentation or product launch. You'll hear how Adolf Hitler's rise to power can be explained by the same type of design flaw that caused the 1999 Mars Polar Lander to crash. You'll also learn how the same simple strategy that saved hundreds of thousands of premature infants also salvaged the Mars Exploration Rovers mission after it was canceled. Finally, I'll share what one of the most misunderstood scientific concepts can teach you about human behavior.

The third and final stage is *achieve*. You'll learn why the final ingredients for unlocking your full potential include both success and failure. You'll discover why the "fail fast, fail often" mantra can be a recipe for disaster. I'll reveal how the same breakdown that led to the collapse of an industry giant also caused a space shuttle to explode. I'll explain why companies pay lip service to learning from failure but don't follow through in practice. We'll discover the surprising benefits of treating success and failure the same and why top performers see uninterrupted success as a warning sign.

By the end of the third stage, instead of letting the world shape your thoughts, you'll let your thoughts shape the world. And instead of simply thinking outside the box, you'll be able to bend the box to your will.

....

THIS IS THE point in the introduction where I'm supposed to lay out a neat personal story about why I'm writing this book. For a book like this, a sensible narrative would involve getting a telescope as a child, falling in love with the stars, pursuing a lifelong career in rocket science, and continuing a passion that culminated in this book—a nice, linear storyline.

But my storyline looks nothing like that. And I won't even try to bend it into a perfect, yet misleading shape. I did get a telescope as a child—well, it was more like a pair of crappy binoculars—but I could never make it work (which should have been a sign). I did have a career in rocket science—until I quit. How I ended up here, as you'll see in the next few pages, is a slapdash combination of good fortune, an excellent mentor, a few good decisions, and perhaps a clerical error or two.

I came to America for all the cliché reasons. When I was a young boy growing up in Istanbul, America assumed a dreamlike quality for me. My vision was culled from the eclectic set of American television shows selected for translation to Turkish. To me, America was Cousin Larry in *Perfect Strangers*, who takes Balki, his Eastern European cousin, under his wing in his Chicago home, where they perform the "dance of joy" to celebrate good fortune. America was *ALF* and the Tanner family, who provides shelter to a furry extraterrestrial with a penchant for trying to eat their cat.

I thought that if America had a place for the likes of Balki and ALF, perhaps it had a place for me, too.

I was born into modest circumstances and wanted better opportunities in life. My dad started working at age six to help support his bus-driver father and homemaker mother. He would wake up before dawn to pick up newspapers hot off the press machines and sell them before school. My mother grew up in rural Turkey, where my grandfather was a shepherd-turned-teacher at a public school. Alongside my grandmother, who was also a teacher, he built the very school they taught in, brick by brick.

When I was growing up, our electricity supply was unreliable and blackouts were a terrifyingly frequent occurrence for a little boy. To keep me distracted, my dad came up with a game. He would light a candle, take my soccer ball, and simulate how the Earth (the soccer ball) rotated around the Sun (the candle).

These were my first astronomy lessons. I was hooked.

At night, I was busy dreaming about the cosmos with half-deflated soccer balls. But during the day, I was a student in a deeply conformist education system. In primary school, our teacher didn't call us Osman or Fatma. Each student was assigned a number, not unlike how livestock are branded for identification purposes. We were 154 or 359 (my number, which I won't disclose, is the only bank PIN I've ever had—"change your PIN frequently" alerts be damned). We wore the same outfits to school—a bright blue uniform with a crisp white collar—and the boys all had the same buzz cut.

Each school day, we recited the national anthem, followed by the standard student oath, where we vowed to dedicate our existence to the Turkish nation. The message was unmistakable: Subjugate yourself, repress your distinctive qualities, and embrace conformity for the greater good.

The task of enforcing conformity eclipsed all other educational priorities. In fourth grade, I once committed the grave sin of skipping a haircut, which immediately drew the ire of my school principal, a bulldozer of a man better suited to be a prison warden. He spotted my longer-than-standard hairdo during one of his inspections and began breathing like a winded rhinoceros. He grabbed a hair clip from a girl and stuck it in my hair as an act of public shaming—a retribution for nonconformity.

Conformity in the educational system saved us from our worst tendencies, those pesky individualistic ambitions to dream big and devise interesting solutions to complex problems. The students who got ahead weren't the contrarians, the creatives, the trailblazers. Rather, you got ahead by pleasing the authority figures, fostering the type of subservience that would serve you well in the industrial workforce.

This rule-following, elder-respecting, rote-memorizing culture left little room for imagination and creativity. This I had to cultivate on my own, primarily through books. My books were my refuge. I bought all that I could afford, handling them gently when reading them, to make sure I didn't bend the pages or the spine. I would lose myself in fantasy worlds created by Ray Bradbury, Isaac Asimov, and Arthur C. Clarke and live vicariously through their fictional characters. I would devour every astronomy book I could find and plaster my walls with posters of scientists like Einstein. On old Betamax tapes, Carl Sagan would speak to me through the original *Cosmos* series. I wasn't quite sure what he was saying, but I listened anyway.

I taught myself how to code and built a website called Space Labs, a digital love letter to astronomy. I would write, in broken, elementary English, all I knew about space. Although my coding skills didn't help me get dates, they would prove to be crucial later in life.

To me, rocket science became synonymous with escape. In Turkey, my path was predetermined. In America—the frontier of rocket science—the possibilities were infinite.

At seventeen, I achieved escape velocity. I was admitted to Cornell University, where my childhood hero, Sagan, had once taught as an astronomy professor. I showed up at Cornell with a thick accent, skinny European jeans, and an embarrassing fondness for Bon Jovi.

Shortly before I arrived at Cornell, I researched what the astronomy department was up to. I learned that an astronomy professor, Steve Squyres, was in charge of a NASA-funded project to send a rover to Mars. He had also worked under Sagan as a graduate student. This was too good to be true.

There was no job posting, but I emailed Squyres my résumé and expressed my burning desire to work for him. I had the lowest of expectations—you might say I was livin' on a prayer—but I remembered one of the best pieces of advice I ever received from my father: You can't win the lottery without buying a ticket.

So I bought a ticket. But I had no idea what I was getting myself into. Much to my surprise, Squyres wrote back and invited me in for an interview. Thanks in part to the coding skills I had picked up in high school, I landed a pinch-me-now job as an operations team member of a mission that would send two rovers, named *Spirit* and *Opportunity*, to Mars. I triple-checked the name on my offer letter to make sure it wasn't some terrible clerical mix-up.

Just a few weeks before, I was in Turkey, daydreaming about space. I now had front-row seats to the action. I channeled my inner Balki and performed the dance of joy. For me, the hope that America was supposed to represent—its spirit and its opportunity—was no longer just a cliché.

I remember the first time I walked into the so-called Mars Room on the fourth floor of the Space Sciences Building at Cornell. Plastered all over the walls were schematics, along with photos of the Martian surface. It was a disorderly, windowless place lit by dreary, headache-inducing fluorescent lights. But I loved it.

I had to learn how to think like a rocket scientist—fast. I spent the first few months listening intently to conversations, reading mountains of documents, and trying to decode the meaning of a whole new set of acronyms. In my spare time, I also worked on the Cassini-Huygens mission, which sent a spacecraft to study Saturn and its surroundings.

Over time, my enthusiasm for astrophysics began to wane. I started to feel a strong disconnect between the theory I studied in class and the practicalities of the real world. I've always been more interested in pragmatic applications than theoretical constructs. I loved learning about the thought process that went into rocket science, but not the substance of the math and physics classes I had to take. I was like a baker who loved rolling out dough but didn't like cookies. There were classmates far better at the substance than I was, and I thought the critical-thinking skills I picked up from my experience could be put to more practical use than the rote work of re-proving why E equals mc^2.

Although I continued my work on the Mars and Saturn missions, I began to explore other options. I found myself far more drawn to the physics of society, and I decided to go to law school. My mother was particularly glad she would no longer have to correct her friends for asking her astrologist son to interpret their horoscopes.

Even after I changed trajectories, I brought with me the tool kit I had acquired from four years of astrophysics. Using the same critical-thinking skills, I graduated first in my law-school class, with the highest grade point average in the law school's history. After graduation, I landed a coveted clerkship on the US Court of Appeals for the Ninth Circuit and practiced law for two years.

I then decided to go into academia. I wanted to bring the insights on critical thinking and creativity that I had obtained from rocket science to education. Inspired by my frustrations with the conformist education system in Turkey, I hoped to empower my students to dream big, challenge assumptions, and actively shape a rapidly evolving world.

Realizing that my reach in the classroom was limited to the enrolled students, I launched an online platform to share these insights with the rest of the world. In my weekly articles, which have reached millions, I write about challenging conventional wisdom and reimagining the status quo.

The truth is, I had no idea where I was heading until I arrived there. Looking back now, I realize that the ending was there from the beginning. A common thread has been there all along, infallibly working itself out across my diverse pursuits. As I meandered from rocket science to law and then to writing and speaking to different audiences, my overarching goal has been to develop a set of tools for thinking like a rocket scientist and to share what I've learned with others. Translating elusive concepts to plain language often requires someone on the outside looking in—someone who knows how rocket scientists think, who can dissect their process, but who is sufficiently removed from that world.

I now find myself sitting on that boundary between insider and outsider, realizing that I accidentally spent my entire life getting ready to write this book.

....

I'M WRITING THESE words at a time when divisions in the world have reached a fever pitch. Despite these earthly conflicts, from a rocket-science perspective, we have more that unites us than divides us. When you look at the Earth from outer space—a blue-and-white interruption in the all-black universe—all earthly boundaries disappear. Every living thing on Earth bears traces of the big bang. As the Roman poet Lucretius wrote, "We are all sprung from celestial seed." Every person on Earth is "gravitationally held on the same 12,742-kilometer-wide wet rock hurtling through space," explains Bill Nye. "There's no option to go it alone. We are all on this ride together."[16]

The vastness of the universe puts our earthly concerns in proper context. It unites us by a common human spirit—one that has gazed up at the same night sky over the millennia, seeing for trillions of miles into the stars, looking thousands of years back in time, and posing the same questions: Who are we? Where do we come from? And where are we going?

The *Voyager 1* spacecraft took off in 1977 to paint the first portrait of the outer solar system, photographing Jupiter, Saturn, and beyond. When it completed its mission at the fringes of our solar system, Sagan came up with the idea of turning its cameras around and pointing them at Earth to take one final image. The now-iconic photo, known as the Pale Blue Dot, depicts Earth as a tiny pixel—a barely perceptible "mote of dust suspended in a sunbeam," in Sagan's memorable words.[17]

We tend to see ourselves at the center of everything. But from the vantage of outer space, the Earth is "a lonely speck in the great enveloping cosmic dark." Reflecting on the deeper meaning of the Pale Blue Dot, Sagan said, "Think of the rivers of blood spilled by all those generals and emperors so that, in glory and triumph, they could become the momentary masters of a fraction of a dot. Think of the endless cruelties visited by the inhabitants of one corner of this pixel on the scarcely distinguishable inhabitants of some other corner."

Rocket science teaches us about our limited role in the cosmos and reminds us to be gentler and kinder to one another. We're in this life for a momentary blip, making the briefest of stands. Let's make that brief stand count.

When you learn how to think like a rocket scientist, you won't just change the way you view the world. You'll be empowered to change the world itself.

STAGE ONE
LAUNCH

In this first stage of the book, you'll learn how to harness the power of uncertainty, reason from first principles, ignite breakthroughs with thought experiments, and employ moonshot thinking to transform your life and business.

1

FLYING IN THE FACE
OF UNCERTAINTY

The Superpower of Doubt

Genius hesitates.

—CARLO ROVELLI

ROUGHLY SIXTEEN MILLION years ago, a giant asteroid is believed to have collided with the Martian surface. That collision dislodged a piece of rock and launched it on a journey from Mars to Earth. The rock landed in Allan Hills in Antarctica thirteen thousand years ago and was discovered in 1984 during a snowmobile ride. As the first rock to be collected from Allan Hills in 1984, it was given the name ALH 84001. The rock would have been cataloged, studied, and then promptly forgotten— were it not for an astonishing secret that appeared to be embedded within.[1]

For millennia, humankind has pondered the same question: Are we alone in the universe? Our ancestors glanced upward in thought, contemplating whether they were cosmic commoners or outliers. As technology progressed, we listened for signals beamed across the universe hoping to capture a message from another civilization. We sent spacecraft across the solar system searching for signs of life. In each case, we came up short.

Until August 7, 1996.

On that date, scientists revealed that they had found organic molecules of biological origin in ALH 84001. Many media outlets were quick to announce these findings as fact of life on another planet. CBS, for example, reported that scientists had "detected single-cell structures on the meteorite—possibly, tiny fossils, and chemical evidence of past biological activity. In other words, life on Mars."[2] CNN's early reports quoted a NASA source who said these structures looked like "little maggots," suggesting they were the remains of complex organisms.[3] The media deluge generated existential hysteria across the globe, prompting then President Clinton to give a major public address on the discovery.[4]

But there was a slight problem. The evidence wasn't conclusive. The scientific paper that formed the basis for these headlines was candid about its inherent uncertainties. Part of its title was "*Possible* Relic Biogenic Activity in Martian Meteorite ALH84001" (emphasis mine).[5] The abstract expressly noted that the features observed on the meteorite "could thus be fossil remains of a past martian biota" but underscored that "inorganic formation is possible." In other words, the molecules may have been the products not of Martian bacteria but of nonbiological activity (e.g., a geological process like erosion). The paper concluded that the evidence is merely "compatible" with life.

But these nuances were glossed over in many of the secondhand translations provided to the public by the media. The incident became infamous, prompting Dan Brown to pen a novel, *Deception Point*, about a conspiracy surrounding extraterrestrial life found on a Martian meteorite.

Everything turned out for the best—at least from the perspective of a book chapter on uncertainty. More than two decades later, the uncertainty lingers. Researchers continue to debate whether Martian bacteria or inorganic activity is responsible for the molecules observed on the meteorite.[6]

It would be tempting to say the media got it wrong, but that would be the same kind of overstatement that dominated the original press coverage of the meteorite. More accurately, we can say that people made a classic mistake: trying to make something appear definite when in fact it isn't.

This chapter is about how to stop fighting uncertainty and harness its power. You'll learn how our obsession with certainty leads us astray and why all progress takes place in uncertain conditions. I'll reveal Einstein's biggest mistake regarding uncertainty and discuss what you can learn from the solution to a centuries-old math mystery. You'll discover why rocket science resembles a high-stakes game of peekaboo, what you can learn from Pluto's demotion as a planet, and why NASA engineers religiously munch on peanuts during critical events. I'll end the chapter with strategies that rocket scientists and astronauts use to manage uncertainty and explain how you can apply them in your own life.

The Certainty Fetish

The Jet Propulsion Laboratory, known as JPL, is a small city of scientists and engineers in Pasadena, California. Located just east of Hollywood, JPL has been responsible for operating interplanetary spacecraft for decades. If you've ever seen video footage of a Mars landing, you've seen the inside of JPL's mission support area.

During a typical Mars landing, the area is packed with row after row of overcaffeinated scientists and engineers eating bags of peanuts and staring at the data pouring into their consoles, while giving the audience the illusion that they are in control. But they are not in control. They're simply reporting the events as a sports announcer might—albeit with fancier language like "cruise stage separation" and "heat shield deploy." They're spectators to a game that ended twelve minutes ago on Mars, and they have no idea what the score is yet.

On average, it takes roughly twelve minutes for a signal from Mars to reach Earth traveling at the speed of light.[7] If something is wrong, and a scientist on Earth spots and responds to the problem in a split second, another twelve minutes will pass for that command to reach Mars. That's twenty-four minutes round trip, but it takes about six minutes for a spacecraft to descend from the top of the Martian atmosphere down to the surface. All we can do is load up the spacecraft with instructions ahead of time and put Sir Isaac Newton in the driver's seat.

That's where the peanuts come in. In the early 1960s, JPL was in charge of the unmanned Ranger missions, which were designed to study the Moon to pave the way for the Apollo astronauts. The Ranger spacecraft would be launched toward the Moon, take close-up photos of the lunar surface, and beam those images back to Earth before plummeting into the Moon.[8] The first six missions ended in failure, leading critics to accuse JPL officials of adopting a cavalier "shoot-and-hope" approach.[9] But a later mission succeeded when a JPL engineer happened to bring peanuts to the mission control room. From then on, peanuts became a staple at JPL for each landing.

In critical moments, these otherwise rational, no-nonsense rocket scientists—who have dedicated their lives to exploring the unknown—look for certainty at the bottom of a Planters peanut bag. As if that's not enough, many of them wear their worn-out good-luck jeans or bring a talisman from a previous successful landing—doing everything that a dedicated sports fan might do to create the illusion of certainty and control.[10]

If the landing goes successfully, Mission Control promptly morphs into a circus. There's no trace of cool and calm. Instead, having conquered the beast of uncertainty, engineers will begin jumping up and down, high-fiving, fist pumping, bear hugging, and disappearing into puddles of joyful tears.

We're all programmed with the same fear of the uncertain. Our predecessors who weren't afraid of the unknown became food for saber-toothed tigers. But the ancestors who viewed uncertainty as life-threatening lived long enough to pass their genes on to us.

In the modern world, we look for certainty in uncertain places. We search for order in chaos, the right answer in ambiguity, and conviction in complexity. "We spend far more time and effort on trying to control the world," Yuval Noah Harari writes, "than on trying to understand it."[11] We look for the step-by-step formula, the shortcut, the hack—the right bag of peanuts. Over time, we lose our ability to interact with the unknown.

Our approach reminds me of the classic story of the drunk man searching for his keys under a street lamp at night. He knows he lost his keys somewhere on the dark side of the street but looks for them underneath the lamp, because that's where the light is.

Our yearning for certainty leads us to pursue seemingly safe solutions—by looking for our keys under street lamps. Instead of taking the risky walk into the dark, we stay within our current state, however inferior it may be. Marketers use the same bag of tricks over and over again but expect different results. Aspiring entrepreneurs remain in dead-end jobs because of the certainty they get in the form of a seemingly stable paycheck. Pharma companies develop me-too drugs that offer only marginal improvement over the competition as opposed to developing the one that's going to cure Alzheimer's disease.

But it's only when we sacrifice the certainty of answers, when we take our training wheels off, and when we dare to wander away from the street lamps that breakthroughs happen. If you stick to the familiar, you won't find the unexpected. Those who get ahead in this century will dance with the great unknown and find danger, rather than comfort, in the status quo.

The Great Unknown

In the seventeenth century, Pierre de Fermat scribbled a note on a textbook margin that would baffle mathematicians for more than three centuries.[12]

Fermat had a theory. He proposed that there's no solution to the formula $a^n + b^n = c^n$ for any n greater than 2. "I have a truly marvelous demonstration of this proposition," he wrote, "which this margin is too narrow to contain." And that's all he wrote.

Fermat died before supplying the missing proof for what came to be known as Fermat's last theorem. The teaser he left behind continued to tantalize mathematicians for centuries (and made them wish Fermat had a bigger book to write on). Generations of mathematicians tried—and failed—to prove Fermat's last theorem.

Until Andrew Wiles came along.

For most ten-year-olds, the definition of a good time doesn't include reading math books for fun. But Wiles was no ordinary ten-year-old. He would hang out at his local library in Cambridge, England, and surf the shelves for math books.

One day, he spotted a book devoted entirely to Fermat's last theorem. He was mesmerized by the mystery of a theorem that was so easy to state, yet so difficult to prove. Lacking the mathematical chops to tackle the proof, he set it aside for over two decades.

He returned to the theorem later in life as a math professor and devoted seven years to working on it in secrecy. In an ambiguously titled 1993 lecture in Cambridge, Wiles publicly revealed that he had solved the centuries-old mystery of Fermat's last theorem. The announcement sent mathematicians into a tizzy: "It's the most exciting thing that's happened in—geez—maybe ever, in mathematics," said Leonard Adleman, professor of computer science at the University of Southern California and Turing Award winner. Even the *New York Times* ran a front-page story on the discovery, exclaiming, "At Last, Shout of 'Eureka!' in Age-Old Math Mystery."[13]

But the celebrations proved premature. Wiles had made a mistake in a critical part of his proof. The mistake emerged during the peer-review process after Wiles submitted his proof for publication. It would take another year, and collaboration with another mathematician, to repair the proof.

Reflecting on how he eventually managed to prove the theorem, Wiles compared the process of discovery to navigating a dark mansion. You start in the first room, he said, and spend months groping, poking, and bumping into things. After tremendous disorientation and confusion, you might eventually find the light switch. You then move on to the next dark room and begin all over again. These breakthroughs, Wiles explained, are "the culmination of—and couldn't exist without—the many months of stumbling around in the dark that [precede] them."

Einstein described his own discovery process in similar terms: "Our final results appear almost self-evident," he said, "but the years of searching in the dark for a truth that one feels, but cannot express; the intense desire and the alternations of confidence and misgiving, until one breaks through to clarity and understanding, are only known to him who has himself experienced them."[14]

In some cases, scientists keep stumbling around in the dark room, and the search continues well past their lifetime. Even when they find the light switch, it may illuminate only part of the room, revealing that the remainder is far bigger—and far darker—than they imagined. But to them, stumbling around in the dark is far more interesting than sitting outside in well-lit corridors.

In school, we're given the false impression that scientists took a straight path to the light switch. There's one curriculum, one right way to study science, and one right formula that spits out the correct answer on a standardized test. Textbooks with lofty titles like *The Principles of Physics* magically reveal "the principles" in three hundred pages. An authority figure then steps up to the lectern to feed us "the truth." Textbooks, explained theoretical physicist David Gross in his Nobel lecture, "often ignore the many alternate paths that people wandered down, the many false clues they followed, the many misconceptions they had."[15] We learn about Newton's "laws"—as if they arrived by a grand divine visitation or a stroke of genius—but not the years he spent exploring, revising, and tweaking them. The laws that Newton failed to establish—most notably his experiments in alchemy, which attempted, and spectacularly failed, to turn lead into gold—don't make the cut as part of the one-dimensional story told in physics classrooms. Instead, our education system turns the life stories of these scientists from lead to gold.

As adults, we fail to outgrow this conditioning. We believe (or pretend to believe) there is one right answer to each question. We believe that this right answer has already been discovered by someone far smarter than us. We believe the answer can therefore be found in a Google search, acquired from the latest "3 Hacks to More Happiness" article, or handed to us from a self-proclaimed life coach.

Here's the problem: Answers are no longer a scarce commodity, and knowledge has never been cheaper. By the time we've figured out the facts—by the time Google, Alexa, or Siri can spit out the answer—the world has moved on.

Obviously, answers aren't irrelevant. You must know some answers before you can begin asking the right questions. But the answers simply serve as a launch pad to discovery. They're the beginning, not the end.

Be careful if you spend your days finding right answers by following a straight path to the light switch. If the drugs you're developing were certain to work, if your client were certain to be acquitted in court, or if your Mars rover were certain to land, your jobs wouldn't exist.

Our ability to make the most out of uncertainty is what creates the most potential value. We should be fueled not by a desire for a quick catharsis but by intrigue. Where certainty ends, progress begins.

Our obsession with certainty has another side effect. It distorts our vision through a set of funhouse mirrors called unknown knowns.

Unknown Knowns

On February 12, 2002, amid escalating tensions between the United States and Iraq, US secretary of defense Donald Rumsfeld took the stage at a press briefing. He received a question from a reporter about whether there was any evidence of Iraqi weapons of mass destruction—the basis for the subsequent American invasion. A typical answer would be packaged in preapproved political stock phrases like *ongoing investigation* and *national security*. But Rumsfeld instead pulled out a rocket-science metaphor from his linguistic grab bag: "There are known knowns; there are things we know we know. We also know there are known unknowns; that is to say we know there are some things we do not know. But there are also unknown unknowns—the ones we don't know we don't know."[16]

These remarks were widely ridiculed—in part because of their controversial source—but as far as political statements go, they're surprisingly accurate. In his autobiography, *Known and Unknown*, Rumsfeld acknowledges that he first heard the terms from NASA administrator William Graham.[17] But Rumsfeld conspicuously omitted one category from his speech—unknown knowns.

Anosognosic is an unpronounceable word used to describe someone with a medical condition that makes them unaware they're suffering from it. For example, if you put a pencil in front of a paralyzed anosognosic individual and ask them to pick it up, they won't do it. If you ask them why, they'll respond, "'Well, I'm tired,' or 'I don't need a pencil.'" As psychologist David Dunning explains, "They literally aren't alerted to their own paralysis."[18]

The unknown knowns are like anosognosia. This is the land of self-delusion. In this category, we think we know what we know, but we don't. We assume we have a lock on the truth—that the ground underneath our feet is stable—but we're actually standing on a fragile platform that can tumble over with a rogue gust of wind.

We find ourselves on that fragile platform far more often than we realize. In our certainty-obsessed public discourse, we avoid reckoning with nuances. The resulting public discussion operates without a rigorous system for discerning proven facts from best guesses. A lot of what we know simply isn't accurate, and it's not always easy to recognize which part lacks real evidence. We've mastered the art of pretending to have an opinion—smiling, nodding, and bluffing our way through a makeshift answer. We've been told to "fake it until we make it," and we've become experts at the faking part. We value chest beating and delivering clear answers with conviction, even when we have little more than two minutes of Wikipedia knowledge on an issue. We march on, pretending to know what we think we know, oblivious to glaring facts that contradict our ironclad beliefs.

"The great obstacle to discovering," historian Daniel J. Boorstin writes, "was not ignorance but the illusion of knowledge."[19] The pretense of knowledge closes our ears and shuts off incoming educational signals from outside sources. Certainty blinds us to our own paralysis. The more we speak our version of the truth, preferably with passion and exaggerated

hand gestures, the more our egos inflate to the size of skyscrapers, concealing what's underneath.

Ego and hubris are part of the problem. The other part is the human
distaste for uncertainty. Nature, as Aristotle said, abhors a vacuum. He
argued that a vacuum, once formed, would be filled by the dense material surrounding it. Aristotle's principle applies well beyond the realm of
physics. When there's a vacuum of understanding—when we're operating
in the land of unknowns and uncertainty—myths and stories whoosh in
to fill the gap. "We can't live in a state of perpetual doubt," Nobel Prize–
winning psychologist Daniel Kahneman explains, "so we make up the best
story possible and we live as if this story were true."[20]

Stories provide the perfect remedy for our fear of uncertainty. They
fill the gaps in our understanding. They create order out of chaos, clarity out of complexity, and a cause-and-effect relationship out of coin-
cidence. Your child exhibits signs of autism? Blame it on that vaccine
the kid got two weeks ago. You spotted a human face on Mars? Must
be the elaborate work of an ancient civilization that, coincidentally, also
helped the Egyptians build the pyramids of Giza. People got sick and
died in clusters, with some of the corpses twitching or making noises?
Vampires, our predecessors concluded, before we knew about viruses and
rigor mortis.[21]

When we prefer the seeming stability of stories to the messy reality of
uncertainty, facts become dispensable and misinformation thrives. Fake
news is not a modern phenomenon. Between a good story and a bunch
of data, the story has always prevailed. These mentally vivid images strike
a deep, lasting chord known as the narrative fallacy. We remember what
so-and-so told us about how his male-pattern baldness was caused by too
much time in the sun. We fall for the story, throwing logic and skepticism
to the wind.

Authorities then turn these stories into sacred truths. All the facts in
the world can't keep democratically elected hate machines from taking
office as long as they can inject a false sense of certainty into an inherently
uncertain world. Confident conclusions by loud-mouthed demagogues

who pride themselves on rejecting critical thinking begin to dominate the public discourse.

What they lack in knowledge, the demagogues make up for by cranking up their assertiveness. As viewers sag in confusion trying to interpret the unfolding facts, the firebrands provide us comfort. They don't bother us with ambiguity or let nuances get in the way of bumper-sticker sound bites. We put our mouths on the spigot of their seemingly clear opinions, happily removing the burden of critical thinking from our shoulders.

The problem with the modern world, as Bertrand Russell put it, is that "the stupid are cocksure while the intelligent are full of doubt." Even after physicist Richard Feynman earned a Nobel prize, he thought of himself as a "confused ape" and approached everything around him with the same level of curiosity, which enabled him to see nuances that others dismissed. "I think it's much more interesting to live not knowing," he remarked, "than to have answers which might be wrong."

Feynman's mindset requires an admission of ignorance and a good dose of humility. When we utter those three dreaded words—*I don't know*—our ego deflates, our mind opens, and our ears perk up. Admitting ignorance doesn't mean remaining willfully oblivious to facts. Rather, it requires a conscious type of uncertainty where you become fully aware of what you don't know in order to learn and grow.

Yes, this approach may illuminate things you don't want to see. But it's far better to be uncomfortably uncertain than comfortably wrong. In the end, it's the confused apes—the connoisseurs of uncertainty—that transform the world.

Connoisseurs of Uncertainty

"Something unknown is doing we don't know what—that is what our theory amounts to."[22]

This is how the astrophysicist Arthur Eddington described the state of quantum theory in 1929. He may as well have been speaking about our understanding of the entire universe.

Astronomers live and work in a dark mansion that's only 5 percent lit. Roughly 95 percent of the universe is made up of ominous-sounding stuff called dark matter and dark energy.[23] They don't interact with light, so we can't see or otherwise detect them. We know nothing about their nature. But we know that they're there because they exert gravitational force on other objects.[24]

"Thoroughly conscious ignorance," physicist James Maxwell said, "is the prelude to any real advance in knowledge."[25] Astronomers reach beyond the borders of knowledge and take a quantum leap into a vast ocean of unknowns. They know that the universe is like a giant onion, where the unwrapping of one layer of mystery simply reveals another. Science, as George Bernard Shaw said, "can never solve one problem without raising 10 more problems."[26] As some gaps in our knowledge are filled, others emerge.

Einstein described this dance with mystery as "the most beautiful experience."[27] Scientists stand "at the edge between known and unknown," physicist Alan Lightman writes, "and gaze into that cavern and be exhilarated rather than frightened."[28] Instead of freaking out over their collective ignorance, they thrive on it. The uncertain becomes a call to action.

Steve Squyres is a connoisseur of uncertainty. He was the principal investigator of the Mars Exploration Rovers project when I served on the operations team. The intensity of his passion for the unknown is contagious. The fourth floor of the Space Sciences Building at Cornell University, where Squyres's office is located, would buzz with energy whenever the doctor was in. When talk turned to Mars (which was often), his eyes glinted with a fiery passion. Squyres is a natural leader. When he moves, others follow. And like any good leader, he's quick to take the blame but also share the credit. He once crossed out his name on an award he had received for his work on a mission, wrote in the names of the staff members who did the heavy lifting, and gave it to them.

Squyres was born in southern New Jersey and inherited his enthusiasm for exploration from his scientist parents.[29] Nothing flared his imagination like the unknown. "When I was a kid," Squyres recalls, "we had an atlas in our home that was fifteen or twenty years old, and there were places where there wasn't a whole lot drawn. I always thought that the idea of a map that had blank spots on it that needed to be filled in was incredibly cool." He dedicated the rest of his life to finding and filling those blank spots.

As an undergraduate at Cornell, he took a graduate-level astronomy course taught by a professor serving on the science team for the Viking mission that sent two probes to Mars. The course required Squyres to write an original term paper. For inspiration, he walked into a room on campus where images of Mars taken by the Viking orbiters were collecting dust. He planned to spend fifteen or twenty minutes looking through photos. "I walked out of that room four hours later," Squyres explains, "knowing exactly what I wanted to do for the rest of my life."

He had found the blank canvas he was looking for. Long after he left the building, his mind continued to hum with the images of the Martian surface. "I didn't understand what I was looking at in these pictures," Squyres says, "but the beauty of it was, nobody did. That was what appealed to me."

The appeal of the unknown led Squyres to become an astronomy professor at Cornell. Even after more than three decades navigating the unknown, "I still haven't gotten over that rush," he says, "that feeling of excitement that comes from seeing something that nobody's ever seen before."

But it's not just astronomers who relish the unknown. Take it from another Steve. At the beginning of each movie scene, Steven Spielberg finds himself surrounded by enormous uncertainty. "Every time I start a new scene, I'm nervous," he explains. "I don't know what I'm gonna think of hearing the lines, I don't know what I'm gonna tell the actors, I don't know where I'm gonna put the camera."[30] Placed in the same situation, others might panic, but Spielberg describes it as "the greatest feeling in the world." He knows that only conditions of tremendous uncertainty bring out his creative best.

All progress—in rocket science, in movies, in your fill-in-the-blank enterprise—takes place in dark rooms. Yet most of us are afraid of the dark. Panic begins to set in the moment we abandon the comfort of light. We fill the dark rooms with our worst fears and stockpile goods waiting for the apocalypse to arrive.

But uncertainty rarely produces a mushroom cloud. Uncertainty leads to joy, discovery, and the fulfillment of your full potential. Uncertainty means doing things no one has done before and discovering things that, for at least a brief moment, no other person has seen. Life offers more of itself when we treat uncertainty as a friend, not a foe.

What's more, most dark rooms come with two-way—not one-way—doors. Many of our excursions into the unknown are reversible. As business magnate Richard Branson writes, "You can walk through, see how it feels, and walk back through to the other side if it isn't working."[31] You just have to leave the door unlocked. This was Branson's approach with the launch of his airline, Virgin Atlantic. His deal with Boeing allowed him to return the first plane he bought if the new airline didn't take off. Branson turned what looked like a one-way door into a two-way door—a move that allowed him to walk out if he didn't like what he saw.

Walking, though, isn't the right metaphor. The connoisseurs of uncertainty don't just walk into dark rooms. They dance in them. And I don't mean the awkward, "arms apart" middle school dance, where you maintain a strict one-foot separation from your crush while attempting to make small talk. No, their dance is more like the tango: sleek, intimate, and uncomfortably and beautifully close. They know that the best way to find the light isn't to push uncertainty away, but is to fall straight into its arms.

The connoisseurs of uncertainty know that an experiment with a known outcome is not an experiment at all and that revisiting the same answers is not progress. If we explore only well-trodden paths, if we avoid games we don't know how to play, we'll remain stagnant. Only when you're dancing in the dark, only when you don't know where the light switch is—or even *what* a light switch is—can progress begin.

First chaos, then breakthrough. When the dance stops, so does progress.

A Theory of Everything

Einstein tangoed with uncertainty for most of his life.[32] He conducted imaginative thought experiments, asked questions that no human before had even thought of asking, and unlocked the deepest mysteries of the universe.

Yet, later in his career, he began to look more and more for certainty. He was bothered that we had two sets of laws to explain how the universe works: the theory of relativity for very big objects and quantum mechanics for the very small. He wanted to bring unity to this discordance and create a single, coherent, beautiful set of equations to rule them all: a theory of everything.

The uncertainty of quantum mechanics particularly bothered Einstein. As science writer Jim Baggott explains, "Physics before the quantum had always been about doing *this* and getting *that*," but "the new quantum mechanics appeared to say that when we do *this*, we get *that* only with a certain probability" (even then, in some circumstances, "we might get *the other*").[33] Einstein remained a self-proclaimed "fanatic believer" that a unified theory would resolve the uncertainty and ensure he wouldn't face what he called the "evil quanta."[34]

But the more Einstein grasped for a unified theory, the more the answers eluded him. In searching for certainty, Einstein lost his sense of wonder and the type of open-minded thought experiments that characterized much of his earlier work.[35]

The search for certainty in a world of uncertainty is a human quest. We all long for absolutes, action and reaction, and neat cause-and-effect relationships where A inexorably leads to B. In our approximations and PowerPoint decks, one variable produces one result, in a straight line. There are no curves or fractions to muddy the waters.

But the reality—as is often the case with reality—is far more nuanced. In his earlier years, Einstein used the phrase "it seems to me" in proposing that light is made up of photons.[36] Charles Darwin introduced evolution with "I think."[37] Michael Faraday spoke of the "hesitation" he experienced when he introduced magnetic fields.[38] When Kennedy pledged to put a man on the Moon, he acknowledged that we were taking a leap into the unknown. "This is in some measure an act of faith and vision," he explained to the American public, "for we do not now know what benefits await us."

These statements don't make for great sound bites. But they have the virtue of being more likely to be correct.

"Scientific knowledge," Feynman explains, "is a body of statements of varying degrees of certainty—some most unsure, some nearly sure, none *absolutely* certain."[39] When scientists make statements, "the question is not whether it is true or false but rather how likely it is to be true or false." In science, absolutes are rejected in favor of a spectrum, and uncertainty is institutionalized. Scientific answers appear in the form of approximations and models, bathed in mystery and complexity. There are margins of error and confidence intervals. What's reported as fact—as in the case of the Martian meteor—is often just a probability.

I find it comforting that there isn't a theory of everything, *the* definitive answer to every question asked. The theories and the paths are multiple. There's more than one right way to land on Mars, more than one right way to organize this book (as I keep telling myself), or more than one right strategy for scaling your business.

In looking for certainty, Einstein got in his own way. But his quest for a theory of everything may also have been ahead of his time. Today, many scientists picked up the baton and continue Einstein's quest for a central idea that unites our understanding of physical laws. Some of these efforts are promising, but they haven't yet borne fruit. Any future breakthroughs will occur only when scientists embrace uncertainty and pay close attention to one of the primary drivers of progress: anomalies.

That's Funny

William Herschel was an eighteenth-century German-born composer who later emigrated to England.[40] He quickly established himself as a versatile musician who could play the piano, the cello, and the violin, going on to compose twenty-four symphonies. But it was another composition—of a nonmusical kind—that would overshadow Herschel's music career.

Herschel was fascinated with math. Lacking a university education, he turned to books for answers. He devoured volumes on trigonometry, optics, mechanics—and my favorite, James Ferguson's *Astronomy Explained Upon Sir Isaac Newton's Principles, and Made Easy to Those Who Have Not Studied Mathematics.* This was the eighteenth-century version of *Astronomy for Dummies.*

He read books on how to construct telescopes and asked a local mirror-builder to teach him how to build one. Herschel began making telescopes, grinding mirrors for sixteen hours a day and making molds out of manure and straw.

On March 13, 1781, Herschel was in his backyard peering through his homemade telescope and searching the sky for double stars, which are stars that appear close to each other. He spotted in the constellation of Taurus, near its border with Gemini, a peculiar object that seemed out of place. Intrigued by the anomaly, Herschel pointed his telescope at the object again a few nights later and noticed that it had moved against the background stars. "It is a comet," he wrote, "for it has changed its place."[41]

But Herschel's initial hunch was wrong. The object couldn't be a comet. It had no tail. It also failed to follow a typical comet's elliptical orbit.

At the time, Saturn was thought to mark the outer boundary of planets in the solar system. Scientists believed planets didn't exist beyond Saturn. But Herschel's discovery proved the establishment wrong. It turned on a new light switch at the end of the known solar system and doubled its size. Herschel's "comet" turned out to be a new planet that would later be called Uranus, after the god of the sky.

Uranus proved to be an unruly planet. It would erratically speed up and then slow down. It refused to cooperate with Newton's laws of gravity, which accurately predicted motion everywhere from objects here on Earth to the trajectories of planets in space.[42]

This anomaly led the French mathematician Urbain Le Verrier to speculate about the existence of another planet located beyond Saturn. This planet, Le Verrier surmised, might be tugging at Uranus and, depending on their respective locations, either pulling Uranus forward and speeding it up or pulling it back and slowing it down. Using only math—with just "the point of his pen," as Le Verrier's contemporary François Arago put it—Le Verrier found another planet. This new planet, Neptune, was later observed within one degree of where Le Verrier predicted it would be.[43] The astonishing match was produced by a set of laws written by Newton nearly 160 years before.

With Neptune's discovery, it appeared that Newton's laws reigned supreme even at the outer edges of the solar system. Yet there seemed to be a problem with a planet closer to home—Mercury. The planet refused to conform to expectations, deviating from the orbit predicted by Newton's laws. It would have been easy to dismiss this flaw as an aberration—an exception that proves the rule—particularly since Mercury seemed to be the only planet where Newton's laws came up short, and even then, only slightly short.

But this minor anomaly concealed a major flaw with Newton's laws. Einstein seized on the glitch to come up with a new theory that accurately predicted Mercury's orbit. In describing gravity, Newton relied on a rough model that said "things attract each other."[44] Einstein's model, in contrast, was more complex: "Stuff warps space and time."[45] To understand what Einstein meant, imagine putting a bowling ball and some billiard balls on a trampoline.[46] The heavy bowling ball would curve the fabric of the trampoline, causing the lighter billiard balls to move toward it. According to Einstein, gravity worked the same way: It warped the fabric of space and time. The closer you are to the massive bowling ball that is the Sun—and Mercury is the closest planet to the Sun—the stronger the warping of space and time and the greater the deviation from Newton's laws.

The path to the light switch, as these examples show, begins with a switch going off in your own mind when you notice an anomaly. But we're not built to notice anomalies. As children, we're taught to put things into two buckets: good and bad. Brushing your teeth and washing your hands are good. Strangers offering us rides in a sketchy white van are bad. As T. C. Chamberlin writes, "From the good the child expects nothing but good; from the bad, nothing but bad. To expect a good act from the bad, or a bad act from the good, is radically at variance with childhood's mental methods."[47] We believe that, as Asimov describes, "everything that isn't perfectly and completely right is totally and equally wrong."[48]

This oversimplification helps us make sense of the world as children. But as we mature, we fail to outgrow this misleading theory. We go around trying to fit square pegs into round holes and pigeonholing things—and people—into neat categories to create the satisfying, but misleading, illusion of having restored order to a disorderly world.

Anomalies distort this clean picture of good and bad and right and wrong. Life is taxing enough without uncertainty, so we eliminate the uncertainty by ignoring the anomaly. We convince ourselves the anomaly must be an extreme outlier or a measurement error, so we pretend it doesn't exist.

This attitude comes at a huge cost. "Discovery comes not when something goes right," physicist and philosopher Thomas Kuhn explains, "but when something is awry, a novelty that runs counter to what was expected."[49] Asimov famously disputed that "Eureka!" is the most exciting phrase in science. Rather, he observed, scientific development often begins by someone noticing an anomaly and saying, "That's funny . . ."[50] The discovery of quantum mechanics, X-rays, DNA, oxygen, penicillin, and others, all occurred when the scientists embraced, rather than disregarded, anomalies.[51]

Einstein's younger son, Eduard, once asked him why he was famous. In his reply, Einstein cited his ability to spot anomalies that others miss: "When a blind beetle crawls over the surface of a curved branch, it doesn't notice that the track it has covered is indeed curved," he explained, implicitly referring to his theory of relativity. "I was lucky enough to notice what the beetle didn't notice."[52]

But luck, to paraphrase Louis Pasteur, favors the prepared. Only when we pay attention to the subtle clues—there's something off with the data, the explanation seems cursory or superficial, the observation doesn't quite fit the theory—can the old paradigm give way to the new.

As we'll see in the next section, just as the embrace of uncertainty leads to progress, progress itself generates uncertainty, as one discovery calls into question the other.

Getting Plutoed

When it comes to discovering planets, amateur astronomers have a habit of beating the experts to the punch.

In the 1920s, a twenty-year-old Kansas farmer named Clyde Tombaugh was busy building telescopes in his spare time, grinding his lenses and mirrors much like Herschel more than a century before.[53] He would point his homemade telescopes at Mars and Jupiter and make drawings of them. Tombaugh knew that the Lowell Observatory in Arizona was working on planetary astronomy, so on a whim, he sent his drawings to the observatory. The Lowell astronomers were so impressed by Tombaugh's drawings that they offered him a job.

On February 18, 1930, when he was comparing different photographs of the sky, Tombaugh picked up a faint dot shifting back and forth. It turned out to be a planet located beyond Neptune. Located far away from the Sun, the planet was named after the Roman god of the dark underworld: Pluto.

But something was off. The calculations of the newly crowned planet's size kept shrinking. In 1955, astronomers thought that Pluto had a mass similar to that of Earth. Thirteen years later, in 1968, new observations showed Pluto weighing in at roughly 20 percent of the Earth's mass. Pluto continued to shrink until 1978, when calculations decidedly made Pluto a featherweight. Its mass was computed to be only 0.2 percent of the Earth's mass. Pluto had been prematurely declared a planet, even though it was far smaller than the others in its league.

Other developments also began to call Pluto's status into question. Astronomers continued to stumble on round objects beyond Neptune and roughly the same size as Pluto. Yet these were not called planets, simply because Pluto happened to be slightly bigger than them.

The arbitrary benchmark continued to hold until a discovery in October 2003. In that year, a new planet that was believed to be larger than Pluto was discovered. The solar system had a tenth member, located at its outer edge. It was named Eris, after the god of discord and strife.[54]

Eris quickly lived up to its name and began to cause significant amounts of strife. Before Eris's discovery, astronomers hadn't bothered to define the term *planet*, but Eris forced their hands. They had to decide whether Eris was a planet. The task fell to the International Astronomical Union, which designates and categorizes objects in the sky. At a routine meeting in 2006, astronomers voted on the definition of a planet, which both Pluto and Eris failed to meet. With a simple vote, they dethroned Pluto—culture, history, textbooks, Mickey Mouse's dog, and countless planetary mnemonics be damned ("My Very Educated Mother Just Served Us Nine Pizzas" also went out the window).

News coverage made it seem like a group of ill-meaning astronomers aimed a laser beam at everyone's favorite runt planet and shot it out of the sky.[55] Mike Brown, a Caltech professor who led the effort to demote Pluto, didn't help: "Pluto is dead," he declared to the press, with the same gravitas that President Barack Obama had when announcing Osama bin Laden's assassination.[56]

Howls of outrage ensued from thousands of Pluto fans who didn't realize they were Pluto fans until the planet was demoted. Online petitions began to pour in.[57] The American Dialect Society voted *plutoed* as its word of the year in 2006.[58] The word means to "demote or devalue someone or something." A new planetary mnemonic nicely summed up the prevailing popular sentiment: Mean Very Evil Men Just Shortened Up Nature.[59]

Politicians in several states deemed Pluto's demotion worthy of pressing legislative action. The indignant Illinois Senate passed a resolution asserting that Pluto was "unfairly downgraded."[60] The New Mexico House of Representatives opted for more flair, stating that "as Pluto passes overhead through New Mexico's excellent night skies, it [will] be declared a planet."[61]

Pluto was central to the order of the cosmos as we knew it. The finite, unchanging number of planets brought some certainty to the vast uncertainty of the universe. It was something tangible you could teach in school and that teachers could test on standardized exams. Overnight, the universe moved underneath us. If Pluto wasn't a planet—something we had taken for granted for more than seventy years—what else was up for dispute?

These cries of cosmic injustice neglected a crucial fact. Pluto wasn't the first object in our solar system to be demoted, and the backlash against this cosmic demotion wasn't the first of its kind.

No, that honor belonged to our very own planet. When everyone thought that Earth was the center of the cosmic arena, Copernicus came along and demoted Earth to a mere planet with the stroke of his pen. "The motions which seem to us proper to the Sun," Copernicus wrote, "do not arise from it, but from the Earth and our orb, with which we revolve around the sun like any other planet."

Like any other planet. We weren't special. We weren't the center of everything. We were ordinary. Copernicus's discovery, much like Pluto's demotion, shook people's sense of certainty and their place in the universe. As a result, Copernicanism was banished for almost a century.

In Douglas Adams's hilarious book *The Hitchhiker's Guide to the Galaxy*, the supercomputer Deep Thought is asked for the "Answer to the Ultimate Question of Life, the Universe and Everything." After seven and a half million years of deep thought, it spits out a clear, but ultimately meaningless, answer: 42. Although the book's fans have tried to ascribe some symbolic meaning to this number, I think there is none. Adams was simply mocking how humans crave and cling to certainty.

It turned out that the number of planets—nine—was as meaningless as the number 42. For astronomers, this was just another day at the office. Science didn't care about feelings, emotions, or irrational attachments to planets. To be sure, there were dissenters within the astronomical community, but most of them moved on. Logic trumped emotion, a new standard was set, and nine became eight. End of story.

Pluto's assassin, Mike Brown, viewed the planet's demotion as an educational opportunity, rather than a source of resentment. Pluto's story, in his view, would allow teachers to explain why in science, as in life, the path to the right answer is rarely straight.

The origin of the word *planet* makes this clear. *Planet* is derived from a Greek word that means "wanderer." Ancient Greeks looked up at the sky and saw objects that moved against the relatively fixed positions of the stars. They called them wanderers.[62]

Like planets, science wanders. Upheaval precedes progress, and progress generates more upheaval. "People wish to be settled," wrote Ralph Waldo Emerson, but "only as far as they are unsettled is there any hope for them."[63] Those who cling to the past get left behind as the world marches forward.

As the story of Pluto's demotion shows, we tend to respond to uncertainty—no matter how benign—as alarming. But the key to growing comfortable with uncertainty is figuring out what's truly alarming and what's not. And that requires playing a game of peekaboo.

A High-Stakes Game of Peekaboo

Imagine sitting on top of a rocket, with the explosive power of a small nuclear bomb, not knowing whether it will work.

Astronauts call that Tuesday.

The Atlas rocket, which launched the Mercury astronauts into space, was feared as too flimsy. "The Atlas boosters were blowing up every other day down at Cape Canaveral," recalls former astronaut Jim Lovell, who would later become the commander of the ill-fated Apollo 13 mission. "It looked like a very quick way to have a short career. So I took the job."[64] Speaking of the Atlas rocket, Wernher von Braun—a former Nazi who later became a chief architect of the US space program—remarked, "John Glenn is going to ride on that contraption? He should be getting a medal just for sitting on top of it before he takes off."[65] We knew so little about the impact of spaceflight on the human condition that Glenn was instructed to read an eye chart every twenty minutes for fear that weightlessness would distort his vision. If you're wondering what it was like for Glenn to orbit the Earth, it "was like visiting the eye doctor," as author Mary Roach quips.[66]

In pop culture, astronauts like Lovell and Glenn are depicted as a bunch of risk-taking, swaggering hotshots with the guts to breezily sit on top of a dangerous rocket. It makes for good drama, but it misleads. Astronauts maintain their calm not because they have superhuman nerves. It's because they have mastered the art of using knowledge to reduce uncertainty. As astronaut Chris Hadfield explains, "In order to stay calm in a high-stress, high-stakes situation, all you really need is knowledge. . . . Being forced to confront the prospect of failure head-on—to study it, dissect it, tease apart all its components and consequences—really works."[67]

Even when riding on top of a flimsy rocket, many of the early astronauts felt in control because they were personally involved in designing the spacecraft. But they also knew what they didn't know—what to be concerned about and what to ignore. Acknowledging these uncertainties was the first step in resolving them. Once the scientists determined, for example, that they didn't know whether microgravity would mess with eyesight, they asked Glenn to take an eye chart to space with him.

This approach has another upside. If we figure out what we know and what we don't know, we contain uncertainty and reduce the fear associated with it. As author Caroline Webb writes, "The more we place boundaries on the uncertainty . . . the more manageable the remaining ambiguity feels to our brains."[68]

Consider the game of peekaboo. The love of the game is universal: Some version of it is believed to exist in virtually every culture.[69] The language is different, but "the rhythm, dynamics, and shared pleasure" all are the same.[70] A familiar face first appears and then disappears behind someone's hands. The baby sits there, puzzled and slightly alarmed, wondering what's going on. But then the hands are drawn apart, revealing the face and restoring order to the world. Laughter follows.

But laughter doesn't follow—not to the same extent at least—when more uncertainty is introduced.[71] In one study, infants smiled less when a different person appeared instead of the same one. Smiling also decreased when the same person reappeared, but in a different location. Even infants as young as six months old had some expectations of certainty as to the identity and location of the person. When these variables changed unexpectedly, so did the infants' enjoyment.

Knowledge turns an uncertain situation into a high-stakes game of peekaboo. Yes, spaceflight is no laughing matter—there are lives at stake—but astronauts contain uncertainty the same way that infants do: by figuring out who's going to appear on the other end when the hands open up.

The uncertainty we enjoy, whether as infants or as astronauts, is the safe kind. We love safaris from a distance. We love pondering the fate of the characters in *Stranger Things* or reading the latest Stephen King book from the comfort of our couch. The mystery will be resolved, and the killer's mask will drop. But when we don't know who the killer is, when we don't know how the story ends, when the chord is left suspended without the final crescendo—as in *Lost* or *The Sopranos*, both of which ended without a clear summation—our blood begins to boil.

In other words, when uncertainty lacks boundaries, discomfort becomes acute. Letting the amorphous fears of an uncertain future marinate in your head turns up the volume on the drama (all the way to 11). "Fear comes from not knowing what to expect and not feeling you have any control over what's about to happen," writes Hadfield. "When you feel helpless, you're far more afraid than you would be if you knew the facts. If you're not sure what to be alarmed about, everything is alarming."

Determining what to be alarmed about requires following the timeless wisdom of Yoda: "Named must your fear be before banish it you can."[72] The naming, I've found, must be done in writing—with paper and pencil (or pen, if you're into technology). Ask yourself, *What's the worst-case scenario? And how likely is that scenario, given what I know?*

Writing down your concerns and uncertainties—what you know and what you don't know—undresses them. Once you lift up the curtain and turn the unknown unknowns into known unknowns, you defang them. After you see your fears with their masks off, you'll find that the feeling of uncertainty is often far worse than what you fear. You'll also realize that in all likelihood, the things that matter most to you will still be there, no matter what happens.

And don't forget the upside. In addition to considering the worst-case scenario, also ask yourself, *What's the best that can happen?* Our negative thoughts resonate far more than our positive ones do. The brain, to paraphrase psychologist Rick Hanson, is like Velcro for the negative but Teflon for the positive. Unless you consider the best-case scenario along with the worst, your brain will steer you toward the seemingly safest path—inaction. But as a Chinese proverb goes, many a false step was made by standing still. You're more likely to take that first step into the unknown when there's the proverbial pot of gold awaiting at the end.

After you determine what's truly worth being alarmed about, you can take measures to mitigate risks by calling two plays from the rocket-science playbook: redundancies and margins of safety. Let's turn to them now.

Why Redundancies Aren't Redundant

In everyday life, the word *redundancy* has a pejorative meaning.[73] But in rocket science, redundancy can be the difference between success and failure—life and death. Redundancy in aerospace refers to a backup created to avoid a single point of failure that can compromise the entire mission. Spacecraft are designed to operate even when things go wrong—to fail without failing. It's the same reason your car has a spare tire in the back and an emergency brake in the front. If you have a flat or if your brakes malfunction, the substitute picks up the slack.

For example, SpaceX's Falcon 9 rocket has nine engines (as the name implies). These engines are sufficiently isolated from each other so that the spacecraft can complete its mission even if an engine fails.[74] Importantly, the engines are designed to fail gracefully, without compromising other components and endangering the mission. During a Falcon 9 launch in 2012, when one of the engines malfunctioned during flight, the other eight engines kept roaring. The flight computer shut down the defective engine and adjusted the rocket's trajectory to take into account the engine failure. The rocket continued to climb and delivered its cargo to orbit.[75]

Redundancies are also used for the computers on a spacecraft. On Earth, computers crash or freeze all the time, and the odds of failure only increase in the stressful environment of space, filled with vibrations, shocks, changing electrical currents, and fluctuating temperatures.[76] This is why the space shuttle's computers were quad-redundant—meaning there were four computers on board that ran the same software. All four computers would individually "vote" on what to do through a majority-voting system.[77] If one computer failed and began spewing nonsense, it could be outvoted by the other three (yes, folks, rocket science is far more democratic than you imagined).

For redundancies to work, they must function independently. Having four computers on the shuttle sounds wonderful, but since they're all running the same software, a single software bug could simultaneously cripple all four computers. This is why the shuttle also included a fifth backup flight system, loaded with a different software built by a different subcontractor from the other four. If a generic software error crippled the four identical primary computers, the backup system could kick in and return the spacecraft to Earth.

Although redundancy is a good insurance policy, it obeys the law of diminishing returns. After a certain point, piling up additional redundancies unnecessarily increases complexity, weight, and cost. Sure, the Boeing 747 could have twenty-four engines instead of four, but you would have to pay $10,000 to ride in a cramped economy seat from Los Angeles to San Francisco.

Excessive redundancy can also backfire and compromise reliability, instead of improving it. Redundancies add additional points of failure. If the engines on the 747 aren't properly isolated, the explosion of one engine could compromise the others—a risk that increases with each additional engine. This risk prompted Boeing to include only two engines, instead of four, on the 777 after concluding the smaller number would lower the risk of accidents.[78] And as we'll see in a later chapter, the apparent safety that redundancy provides can lead people to make sloppy decisions. They may assume—incorrectly—that even if something goes wrong, there's a fail-safe in place. Redundancy, in other words, can't be a substitute for good design.

Think about it: Where are the redundancies in your own life? Where's the emergency brake or the spare tire in your company? How will you deal with the loss of a valuable team member, a critical distributor, or an important client? What will you do if your household loses a source of income? The system must be designed to continue operating even if a component fails.

Margins of Safety

In addition to including redundancies, rocket scientists address uncertainty by building in margins of safety. For example, they build spacecraft stronger than what appears necessary or make thermal insulation thicker than required. These margins of safety protect the spacecraft in case the uncertain environment of space turns out to be more hostile than expected.

As the stakes go up, so should the margins of safety. Is the probability of failure high? If failure happens, would it be costly? To return to our discussion from earlier, is the door one-way or two-way? If you're making irreversible one-way decisions, go for higher margins of safety.

The decisions we make for spacecraft are mostly irreversible. After the spacecraft is launched, there's no opportunity for a hardware recall. So the tools we include on the spacecraft must be versatile—resembling two-way doors.

Let's go back for a moment to the Mars Exploration Rovers project, which sent two rovers, *Spirit* and *Opportunity*, to the red planet in 2003. There was a tremendous amount of uncertainty about what we would find when we landed. So we adopted a Swiss army knife approach.

In planning the Mars operations, we put a variety of tools on the rovers and made them as flexible and capable as possible. Our rovers had cameras to look around the surface, spectrometers to analyze the composition of the soil and the rocks, a microscopic imager to get close-up views, and a grinding tool that functioned like a hammer to expose the interior of a rock.[79] We could drive our rover—albeit painfully slowly, at about two meters per day—to check out different sites.

At the landing sites for the two rovers, we had some idea of what to expect, having seen snapshots of the regions taken by Martian orbiters.

But our expectations for both landing sites were "totally, completely, utterly wrong," as Steve Squyres puts it.[80] So we learned to use the rovers' tools to solve the problems that Mars gave us—as opposed to the problems we expected.

If the tools on board the spacecraft are sufficiently versatile, they can perform functions that go far beyond their intended use. When *Spirit*'s right front wheel failed in March 2006, the navigators drove the rover backward for the rest of its life.[81] When a mechanical problem crippled the drill on the Martian rover *Curiosity*, engineers invented a new way to drill using the still-functional parts of the rover.[82] After successfully testing the new drilling technique on Earth using a twin rover, they beamed up instructions to *Curiosity* to try it on Mars. It worked beautifully.

The same approach saved the astronauts on the Apollo 13 mission to the Moon. An oxygen tank exploded near the Moon, depleting the power and oxygen supply in the command module. So the three astronauts had to move out of the command module and into the lunar module, using it as a lifeboat to return home. But the lunar module—the small spider-shaped spacecraft designed only to shuttle two astronauts between the lunar surface and the orbiting spacecraft—quickly filled up with dangerous levels of carbon dioxide with three men breathing in and out. There were square canisters in the command module to absorb carbon dioxide, but they wouldn't fit the round filtration system in the lunar module. With help from the ground, astronauts figured out a way—using tube socks and duct tape, among other random objects—to fit a square peg into a round hole.[83]

There are important lessons here for us all. When we face uncertainty, we often manufacture excuses for not getting started. *I'm not qualified. I don't feel ready. I don't have the right contacts. I don't have enough time.* We don't start walking until we find an approach that's guaranteed to work (and preferably one that comes with job satisfaction and a six-figure salary).

But absolute certainty is a mirage. In life, we're required to base our opinions on imperfect information and make a call with sketchy data. "We didn't know what we were doing when we landed" on Mars, Squyres admits. "How can you know what you're doing when no one has done it before?" If our group had postponed until the choices presented themselves with perfect clarity—until we had perfect information about our landing sites so we could design the perfect set of tools for them—we never would have gotten to Mars. Someone else willing to tango with uncertainty would have beaten us to the finish line.

The path, as the mystic poet Rumi writes, won't appear until you start walking. William Herschel started walking, grinding mirrors, and reading astronomy-for-dummies books even though he had no idea he would discover Uranus. Andrew Wiles started walking when he picked up a book on Fermat's last theorem as a teenager, not knowing where his curiosity might lead. Steve Squyres started walking in search of his blank canvas, even though he had no idea it would one day lead him to Mars.

The secret is to start walking before you see a clear path.

Start walking, even though there will be stuck wheels, broken drills, and exploding oxygen tanks ahead.

Start walking because you can learn to walk backward if your wheel gets stuck or you can use duct tape to block catastrophe.

Start walking, and as you become accustomed to walking, watch your fear of dark places disappear.

Start walking because, as Newton's first law goes, objects in motion tend to stay in motion—once you get going, you will keep going.

Start walking because your small steps will eventually become giant leaps.

Start walking, and if it helps, bring a bag of peanuts with you for good luck.

Start walking, not because it's easy, but because it's hard.

Start walking because it's the only way forward.

Visit **ozanvarol.com/rocket** to find worksheets, challenges, and exercises to help you implement the strategies discussed in this chapter.

2

REASONING FROM
FIRST PRINCIPLES

The Ingredient Behind
Every Revolutionary Innovation

Originality consists of returning to the origin.
—ANTONI GAUDÍ

STICKER SHOCK ISN'T in the vocabulary of most Silicon Valley entrepreneurs.

But that's what Elon Musk experienced as he shopped for rockets to send a spacecraft to Mars. On the American market, the price tag for two rockets was a whopping $130 million.[1] The price was for the launch vehicle alone. It didn't include the spacecraft itself, along with its payload, which would further ratchet up the total cost.

So Musk thought he would try his luck in Russia instead. He took several trips there to shop for decommissioned intercontinental ballistic missiles (without the nuclear warheads on top). His vodka-fueled meetings with Russian officials were punctuated by toasts every two minutes (*To space! To America! To America in space!*). But for Musk, the cheers turned into jeers when the Russians told him each missile would set him back

$20 million. As wealthy as Musk was, the cost of the rockets made it too expensive for him to start his space company. He knew he had to do something different.

Since his childhood, the South African has been on a transformational streak, bending one industry after another to his will. At twelve, he programmed and sold his first video game. At seventeen, he immigrated to Canada and later to the United States to major in physics and business at the University of Pennsylvania. He then dropped out of a Stanford PhD program to start a company with his brother, Kimbal. The company, Zip2, was an early provider of online city guides. Too broke to afford an apartment, Elon Musk slept on a futon in his office and showered at the local YMCA.

In 1999, when he was twenty-eight years old, he sold Zip2 to Compaq, instantly becoming a multimillionaire. He then picked up his chips and laid them down on a new table. He harvested his profits from Zip2 to build X.com, an online bank that was later renamed PayPal. When PayPal was acquired by eBay, Musk walked away with $165 million.

Months before the deal was finalized, Musk was already on a beach in Rio de Janeiro. But he wasn't planning his retirement or leafing through the latest Dan Brown novel. No, Musk's idea of beach reading was *Fundamentals of Rocket Propulsion*. The PayPal guy was on a mission to transform himself into the rocket guy.

In its heyday, the space industry was the frontier of innovation. But when Musk thought about entering the business, aerospace companies were hopelessly stuck in the past. Space is the rare tech-related industry that violates Moore's law, the principle named after the Intel cofounder Gordon Moore. According to the principle, computer power develops exponentially, doubling every two years. A computer that would have filled an entire room in the 1970s now fits in your pocket and packs far more computing punch. But rocket technology bucks Moore's law. "We sleep easy knowing that next year's software will be better than this year's," Musk explains, but "rockets' [cost] actually gets progressively worse every year."[2]

Musk wasn't the first to spot this trend. But he was among the first to do something about it.

He launched SpaceX—short for Space Exploration Technologies—with the audacious goal of colonizing Mars and making humanity a multiplanetary species. But Musk's deep pockets weren't enough to buy rockets on the American or the Russian market. He pitched venture capitalists, but they were a hard bunch to convince. "Space is pretty far out of the comfort zone of just about every VC on Earth," Musk explained. He refused to let his friends invest, because he believed the company had only a 10 percent chance of success.

Musk was about to give up when he realized his approach had been deeply flawed. Rather than quit, he decided to go back to first principles—the topic of this chapter.

Before I explain how first-principles thinking works, we'll begin by exploring two obstacles to it. You'll learn why knowledge can be a vice, rather than a virtue, and how a road engineer in the Roman Empire ended up determining the width of NASA's space shuttle. You'll discover the invisible rules that are holding you back and learn how to get rid of them. I'll explain how a pharmaceutical giant and the US military use the same strategy to fend off threats and why killing your business might be the best way to save it. We'll explore why subtracting, rather than adding, is the key to innovation and how a mental model can help simplify your life. You'll walk away from the chapter with practical strategies for putting first-principles thinking to work in your own life.

We've Always Done It This Way

One of my favorite movies, *Animal House*, opens with the camera zooming in on a statue of Emil Faber, the founder of the college where the movie takes place. Inscribed on the statue is a spectacularly banal quote from the fictitious Faber: *Knowledge is good.* The quote is an obvious parody of real-life college founders, who all felt compelled to have an inspiring motto attached to their name. Setting the mockery aside, Faber is undoubtedly correct, and at least in my case, he's preaching to the choir: I make a living as a knowledge worker.

But the same qualities that make knowledge a virtue can also turn it into a vice. Knowledge shapes. Knowledge informs. It creates frameworks, labels, categories, and lenses through which we view the world. It acts as a haze, an Instagram filter, and a poetic structure under which we live our lives. These structures are notoriously hard to beat back, and for good reason: They're useful. They provide us with cognitive shortcuts for making sense of the world. They make us more efficient and productive.

But if we're not careful, they can also distort our vision. If we know, for example, that the market price for rockets is sky-high, we assume that only powerful governments and megacorporations with unique access to large sums of cash can build them. Unwittingly, knowledge can make us a slave to convention. And conventional thoughts lead to conventional results.

When I first started teaching, it struck me as odd that the students at my law school were required to take Criminal Procedure—a difficult class requiring a strong foundation in other topics—in their first year. Over lunch, when I asked a senior colleague to explain, he lowered the newspaper he'd been studying and dismissively remarked, "We've always done it this way." Decades ago, someone decided to structure the curriculum this way, and that was a good enough reason to stick to it. Since then, no one had raised a hand and asked why or why not.

The status quo is a super magnet. People are biased against the way things could be and find comfort in the way things are. If you had any doubts about our obsession with the status quo, take a look at all these idioms we've dedicated to avoiding change: "If it ain't broke, don't fix it." "Don't rock the boat." "Don't change horses in the middle of the stream." "Go with the devil you know."

The default carries immense power, even in advanced industries like rocket science. This idea is called path dependence: What we've done before shapes what we do next.

Here's an example. The width of the engines that powered the space shuttle—one of the most complex machines humankind has ever created—

was determined over two thousand years ago by a Roman road engineer.[3] Yes, you read that correctly. The engines were 4 feet 8.5 inches wide because that was the width of the rail line that would carry them from Utah to Florida. The width of that rail line, in turn, was based on the width of tramlines in England. The width of the tramlines, in turn, was based on the width of the roads built by the Romans: 4 feet 8.5 inches.

The keyboard layout most of us use was designed to be inefficient. Before the current arrangement, typewriters would jam if you typed too quickly. The QWERTY layout (named after the first six letters on the keyboard) was created specifically to slow down typing speed to prevent mechanical key blockage. In addition, for marketing purposes, the letters that make up the word *typewriter* were placed on the top line to allow a salesperson to demonstrate how the machine operates by quickly typing the brand name (try it out!).

Of course, mechanical key blockage is no longer a problem. Nor is there a need to type *typewriter* as quickly as possible. Yet despite the availability of far more efficient and far more ergonomic layouts, the QWERTY arrangement still dominates.

Change can be costly. Abandoning the QWERTY layout for an alternative, for example, would require us to learn to type from scratch (though there's a tribe of people who have made the switch and who argue it's worth the effort). And sometimes things change for the worse. But more often than not, we stick with the default even when the benefits of change far exceed the costs.

Vested interests also reinforce the status quo. High-level executives at *Fortune* 500 companies shun innovation because their compensation is tied to short-term quarterly outcomes that may be temporarily disrupted by forging a new path. "It's difficult to get a man to understand something," Upton Sinclair said, "when his salary depends on his not understanding it."

If you were a horse breeder in Detroit in the early 1900s, you would have assumed that your competition was other breeders raising stronger and faster horses. If you ran a cab company ten years ago, you would have assumed that your competition was other cab companies. If you run airport security, you assume that the primary threat will come from another guy with a bomb in his shoe, so you "solve" terrorism by making everyone take off their shoes.

In each case, the past drowns out the future. Steady as she goes—until you hit an iceberg.

Research shows that we become increasingly rule bound as we grow older.[4] Events begin to rhyme. Days begin to repeat. We regurgitate the same overworn sound bites, stick to the same job, talk to the same people, watch the same shows, and maintain the same product lines. It's a choose-your-own-adventure book that always has the same ending.

The deeper the snow tracks, the harder it is to escape them. An established method of doing things can conceal the exit gate. "When a road is built," wrote Robert Louis Stevenson, "it is a strange thing how it collects traffic, how every year as it goes on, more and more people are found to walk thereon, and others are raised up to repair and perpetuate it, and keep it alive."[5]

We treat our processes and routines like roads collecting traffic. A 2011 survey of more than a hundred American and European companies shows that "over the past 15 years, the amount of procedures, vertical layers, interface structures, coordination bodies, and decision approvals needed in each of those firms has increased by anywhere from 50% to 350%."[6]

Here's the problem. Process, by definition, is backward looking. It was developed in response to yesterday's troubles. If we treat it like a sacred pact—if we don't question it—process can impede forward movement. Over time, our organizational arteries get clogged with outdated procedures.

Complying with these procedures then becomes the benchmark for success. "It's not that rare," Jeff Bezos says, "to hear a junior leader defend a bad outcome with something like, 'Well, we followed the process.'" "If you're not watchful," Bezos warns, "the process can become the thing." But you don't need to throw your standard operating procedures into the shredder and create a corporate free-for-all. Rather, you need to make a habit of asking, as Bezos does, "Do we own the process or does the process own us?"[7]

When necessary, we must unlearn what we know and start over. This is why Andrew Wiles—the mathematician who solved the centuries-old Fermat's last theorem—said, "It's bad to have too good a memory if you want to be a mathematician. You need to forget the way you approached [the problem] the previous time."[8]

In the end, Emil Faber was right. Knowledge *is* good. But knowledge should inform, not constrain. Knowledge should enlighten, not obscure. Only through the evolution of our existing knowledge will the future come into focus.

The tyranny of our knowledge is only part of the problem. We're constrained not only by what we've done in the past, but also by what others have done as well.

They're Doing It This Way

We're genetically programmed to follow the herd. Thousands of years ago, conformity to our tribe was essential to our survival. If we didn't conform, we would be ostracized, rejected, or, worse, left for dead.

In the modern world, most of us yearn to stand out from the herd. We believe we have distinct tastes and a different worldview than does the general population. We might admit interest in other people's choices, but we would argue that our decisions are our own.

The research shows otherwise. In one representative study, participants were quizzed about a documentary they watched: "How many policemen were there when the woman got arrested? What was the color of her dress?"[9] They took the test on their own and didn't see the other participants' responses. A few days later, they returned to the lab to get retested. This time, they were shown the responses of the other participants. But the researchers played a trick: They intentionally doctored some of the responses to be false.

Roughly 70 percent of the time, the participants changed their correct answers and went along with the wrong answers given by the rest of the group. Even after the experimenters told the participants that the group's answers were wrong, the fake social proof was so powerful that about 40 percent of the participants stuck with the wrong answers during retesting.

Resisting this hardwiring for conformity causes us emotional distress—literally. A neurological study showed that nonconformity activates the amygdala and produces what the authors describe as "a pain of independence."[10]

To avoid this pain, we pay lip service to being original, but we become the by-products of other people's behaviors. It's like that Chinese proverb: One dog barks at something, and a hundred others bark at that sound.

Businesses plant their lightning rod where lightning struck last and wait for it to strike again. *This worked once, so let's do it again. And again. And again.* Let's launch the same marketing campaign, use the same formula in that mega-successful mass-market romance book, and make the seventeenth sequel to *Fast and Furious*. Particularly in conditions of uncertainty, we tend to copy and paste from our peers and competitors, assuming they know something we don't.

This strategy can work in the short term, but it's a recipe for long-term disaster. The winds of fashion are fickle, and trends are transitory. Over time, imitation makes the original obsolete. The same path that led to glory for one person can cause catastrophe for another. Conversely, the same path that led to catastrophe for one person can yield glory for another. Friendster and Myspace both fizzled out, yet Facebook's market capitalization was over half a trillion dollars by mid-2019.

To be sure, there's tremendous value in learning what others have mastered. Emulation, after all, is our earliest teacher. Conformity teaches us everything—how to walk, how to tie our shoes, and more. For less than twenty dollars, a book can show you what took someone else a lifetime to figure out. But there's an important difference between learning and blind imitation.

You can't copy and paste someone else's path to success. You can't drop out of Reed College, sit in on a calligraphy class, take some LSD, dabble in Zen Buddhism, set up shop in your parents' garage, and expect to start the next Apple. As Warren Buffett put it, "The five most dangerous words in business are 'Everybody else is doing it.'" This monkey see, monkey do approach creates a race to the exceedingly crowded center—even though there's far less competition on the edges. "When you try to improve on existing techniques," says Astro Teller, the head of X, Google's moonshot factory, "you're in a smartness contest with everyone who came before you. Not a good contest to be in."[11]

Musk initially found himself in this contest when he began shopping for rockets. His thinking was contaminated by what others had done in the past. So he decided to return to his physics training and reason from first principles.

A word about Musk before I proceed. I've found that his name generates unusually strong opinions. Some view him as the real-life Iron Man, the most interesting man in the world, an entrepreneur with a heart who is doing more than anyone else to move humanity forward. Others describe him as a Silicon Valley dilettante whose world-saving companies flirt all too frequently with disaster and a showman who spins self-indulgent tales about the future from his Twitter account (while getting himself into regulatory hot water).

I'm in neither of these camps. I think we do Musk a disservice if we vilify him or fetishize him. But we do ourselves a disservice if we fail to learn from how he has used first-principles thinking to upend numerous industries, turning his starry-eyed dreams into reality.

Back to First Principles

The credit for first-principles thinking goes to Aristotle, who defined it as "the first basis from which a thing is known."[12] The French philosopher and scientist René Descartes described it as systematically doubting everything you can possibly doubt, until you're left with unquestionable truths.[13] Instead of regarding the status quo as an absolute, you take a machete to it. Instead of letting your original vision—or the visions of others—shape the path forward, you abandon all allegiances to them. You hack through existing assumptions as if you're hacking through a jungle until you're left with the fundamental components.

Everything else is negotiable.

First-principles thinking allows you to see the seemingly obvious insight that's hiding under everyone's nose. "Talent hits a target no one else can hit," philosopher Arthur Schopenhauer said, but "genius hits a target no one else can see." When you apply first-principles thinking, you switch from being a cover band that plays someone else's songs to an artist that does the painstaking work of creating something new. You go from what author James Carse calls a finite player, someone playing *within* boundaries, to an infinite player, someone playing *with* boundaries.

Returning home from his last shopping spree in Russia empty-handed, Musk had an epiphany. In trying to buy rockets that others had built, he realized he was playing the role of a cover band—a finite player. On the flight back home, Musk told Jim Cantrell, an aerospace consultant who accompanied Musk on the trip, "I think we can build a rocket ourselves."[14] Musk showed Cantrell a spreadsheet with numbers he had been crunching. Cantrell recalls, "I looked at it and said, I'll be damned—*that's* why he's been borrowing all my [rocketry] books."

"I tend to approach things from a physics framework," Musk explained in a later interview. "Physics teaches you to reason from first principles," he added, "rather than by analogy"—in other words, copying or analogizing from others with little deviation.

For Musk, using first principles meant starting with the laws of physics and asking himself what's required to put a rocket in space. He stripped a rocket down to its smallest subcomponents—its fundamental raw materials. "What is a rocket made of?" he asked himself. "Aerospace-grade aluminum alloys, plus some titanium, copper, and carbon fiber. And then

I asked, what is the value of those materials on the commodity market? It turned out that the materials cost of a rocket was around 2 percent of the typical price—which is a crazy ratio."

The price disparity resulted, at least in part, from a culture of outsourcing in the space industry. Aerospace companies outsource to subcontractors, which then outsource to sub-subcontractors. "You have to go four or five layers down," Musk explained, "to find somebody actually doing something useful—actually cutting metal, shaping atoms."

So Musk decided to cut his own metal and construct his next-generation rockets from scratch. Walk through the halls of SpaceX's factories, and you'll notice people doing everything from welding titanium to building in-flight computers. Roughly 80 percent of all SpaceX rocket components are manufactured in-house. This gives the company greater control over cost, quality, and pace. With few outside vendors, SpaceX can move from idea to execution at record speed.

Here's an example of the benefits of in-house production. Tom Mueller, the propulsion chief at SpaceX, once asked a vendor to build an engine valve. "They said it would cost a quarter million dollars and it would take a year to make," Mueller recalls. He responded, "No, we need it by this summer, for much, much less money." The vendor said, "Good luck with that," and walked away. So Mueller's team built the valve themselves—at a fraction of the cost. When the vendor called Mueller over the summer to ask whether SpaceX still needed the valve, Mueller responded, "We made it, we finished it, we qualified it, and we're going to fly it."[15] Mike Horkachuck, the NASA liaison to SpaceX, was surprised to see how Mueller's approach pervades the entire company: "It was unique because I almost never heard NASA engineers talking about [the] cost of a part when they were making design trades and decisions."[16]

SpaceX also got creative with sourcing raw materials. An employee bought a theodolite, a piece of equipment used to track and align rockets, on eBay for $25,000 after discovering that a new version cost too much. Another employee procured a giant piece of metal from an industrial junkyard to make a fairing—the nose cone that protects the rocket. Cheap, used components, if tested and qualified properly, can work just as well as expensive new ones.

SpaceX also borrowed components from other industries. Instead of using costly equipment to build handles for hatches, the company used parts from bathroom stall latches. Instead of designing expensive custom-built harnesses for astronauts, it used race-car safety belts, which are more comfortable and less expensive. In place of specialty onboard computers that cost up to $1 million, SpaceX's first rocket installed the same type of computer used in an ATM for $5,000. Compared with the total cost of a spacecraft, these cost cutters may not seem like much, but "when you add them all up," Musk says, "it makes a huge difference."

Many of these cheaper components have the benefit of being more reliable. Consider, for example, the fuel injectors used in SpaceX rockets. Most rocket engines use a showerhead design, where multiple injectors spray fuel into a rocket's combustion chamber. SpaceX uses what's called a pintle engine, with only one injector, which looks like the nozzle of a garden hose. The less expensive pintle is also less likely to create combustion instability, which can cause what rocket scientists call a rapid unscheduled disassembly—or what laypeople call an explosion.

First-principles thinking prompted SpaceX to question another deeply held assumption in rocket science.[17] For decades, most rockets that launched spacecraft into outer space couldn't be reused. They would plunge into the ocean or burn up in the atmosphere after carrying their cargo to orbit, requiring an entirely new rocket to be built. It was the cosmic equivalent of torching the airplane at the end of each commercial flight. The cost of a modern rocket is about the same as a Boeing 737, but flying on a 737 is far less expensive because jets, unlike rockets, are flown over and over again.

The solution is obvious: Do the same thing with rockets. This is why parts of NASA's space shuttle were reusable. The solid rocket boosters that carried the shuttle into orbit would separate from the spacecraft and parachute down to the Atlantic Ocean, to be picked up and refurbished.

The orbiter, which carried the astronauts, would also glide back to Earth after each mission to be reused on future flights.

For reusability to make economic sense for rockets, it must be as quick and as complete as possible. In this context, quick means that the reusable parts require minimal postmission investigation and refurbishment. After a rapid inspection and refuel, the rocket should be able to take off—much like an airplane inspected and refueled at the end of a trip. And with complete reusability, all components of the spacecraft are reusable so that no hardware is thrown out.

But for the space shuttle, reusability was neither quick nor complete. The cost of inspection and refurbishment was outrageously high, particularly given the infrequency of shuttle flights. The turnaround required "more than 1.2 million different procedures," taking months and costing more than a new space shuttle.[18]

If you reason by analogy, you would conclude that reusable spacecraft are a bad idea. *It didn't work for NASA, so it won't work for us.* But this reasoning is flawed. The case against reusability was built on a single case study: the space shuttle. The problem, however, was with the shuttle itself, not with all reusable spacecraft.

Rockets come in stages that are stacked on top of each other. SpaceX's Falcon 9 rocket has two stages. The first stage is a fourteen-story-tall portion of the rocket body with nine engines. After the first stage battles gravity and lifts the spacecraft off the launch pad and into space, it separates and drops away, letting the second stage take over. The second stage, which comes with only one engine, ignites and continues to push the spacecraft up. The first stage is the most expensive part of the Falcon 9, representing about 70 percent of the entire cost of the mission. Even if only the first stage can be recovered and reused efficiently, it would save a lot of money.

But recovery and reuse are no easy feats. The first stage must separate from the spacecraft, do a black flip, reignite three of its engines to slow down, find its way to a landing pad on Earth, and gently set its gigantic body upright on the ground. In the words of one SpaceX press release, this feat is like balancing "a rubber broomstick on your hand in the middle of a wind storm."[19]

In December 2015, the first stage of a Falcon 9 rocket successfully completed an upright landing on solid ground after putting its cargo in orbit. Blue Origin—which is Bezos's private spaceflight company—also landed the reusable booster stage of its New Shepard rocket back on Earth after sending it to space. Since then, both companies have refurbished and reused numerous recovered rocket stages, sending them back out to space like certified pre-owned cars. What was once a wild experiment is on its way to becoming routine.

The innovations produced by first-principles thinking enabled Blue Origin and SpaceX to drastically cut the cost of spaceflight. For example, when SpaceX begins to carry NASA astronauts to the International Space Station, each flight is projected to cost $133 million to taxpayers—less than a third of a $450 million space shuttle launch of days past.

SpaceX and Blue Origin had one thing going for them: They were new to the industry. They had the benefit of writing on a blank slate. There were no fixed internal ideas, no long-established practices, and no legacy components. Without the tug of their own past, they could let first principles drive rocket design.

Most of us don't have that luxury. We're inevitably influenced by what we know and the beaten-down paths walked by the pioneers before us. Escaping our own assumptions is tricky business—particularly when they're invisible to us.

How Invisible Rules Hold You Back

Author Elizabeth Gilbert tells the fable of a great saint who would lead his followers in meditation.[20] Just as the followers were dropping into their zen moment, they would be disrupted by a cat that would "walk through the temple meowing and purring and bothering everyone." The saint came up with a simple solution: He began to tie the cat to a pole during medi-tation sessions. This solution quickly developed into a ritual: Tie the cat to the pole first, meditate second.

When the cat eventually died (of natural causes), a religious crisis en-sued. What were the followers supposed to do? How could they possibly meditate without tying the cat to the pole?

This story illustrates what I call invisible rules. These are habits and be-haviors that have unnecessarily rigidified into rules. They're unlike written rules, which are visible. The written rules appear right there in the standard operating procedures and can be amended or deleted.

Although written rules, as we saw above, can be resistant to change, invisible rules are even more stubborn. They're the silent killers that con-strain our thinking without our being aware of it. They turn us into a rat trapped in a Skinner box, pressing the same lever over and over again—except the box was designed by us and we're free to venture out any-time. We're perfectly capable of meditating without the cat, but we don't realize it.

We then make things worse by defending our self-imposed limitations. We could do things differently, we say, but our supply chain, our software, our budget, our skill set, our education, our whatever, doesn't allow it. As the saying goes, argue for your limitations, and you get to keep them.

"Your assumptions are your windows on the world," said Alan Alda, in a quote often misattributed to Asimov. "Scrub them off every once in a while, or the light won't come in."[21] What in your own world is the cat from the meditation fable? Which unnecessary relic of the past clouds your thinking and hampers your progress? What do you assume you're supposed to do simply because everyone around you is doing it? Can you question this assumption and replace it with something better?

We used to assume that a restaurant required tables, an immobile kitchen, and a brick-and-mortar location. Questioning these assumptions gave us food trucks. We used to assume that late fees and physical stores were necessary for video rentals. Questioning these assumptions gave us

Netflix. We used to assume that you needed bank loans or venture-capital funding to launch a new product. Questioning these assumptions gave us Kickstarter and Indiegogo.

To be sure, you can't go through life questioning every single thing you do. Routines free us of the thousands of exhausting daily decisions we would otherwise have to make. For example, I eat the same thing for lunch every day and take the same route to work. I routinely reason by analogy and copy other people's choices when it comes to fashion, music, and interior design (my living room looks like a page out of the Crate & Barrel catalog).

In other words, first-principles thinking should be deployed where it matters the most. To mop the mist collected on your mental windshield in those areas and expose the invisible rules governing your life, spend a day questioning your assumptions. With each commitment, each presumption, each budget item, ask yourself, *What if this weren't true? Why am I doing it this way? Can I get rid of this or replace it with something better?*

Be careful if you find yourself coming up with multiple reasons to keep something. "By invoking more than one reason," observes author and scholar Nassim Nicholas Taleb, "you are trying to convince yourself to do something."[22]

Demand current—not historical—supporting evidence. Many of our invisible rules were developed in response to problems that no longer exist (like the cat in the meditation fable). But the immune response remains long after the pathogen leaves.

The best way to expose invisible rules is to violate them. Go for a seeming moonshot you don't think you'll achieve. Ask for a raise you don't think you deserve. Apply for a job you don't think you'll get.

You'll find, after all, that it *is* possible to meditate without the cat.

First-principles thinking isn't just for finding the fundamental components of a product or practice—whether it's a rocket or your meditation ritual—and building something new. You can also use this thinking to find the raw materials within you and build the new you. This, in turn, requires risking your significance.

Why You Should Risk Your Significance

When Steve Martin first started performing stand-up comedy, there was a proven formula for telling jokes.[23] Each joke came with its own cringe-worthy punchline. Here's a rocket-science example:

QUESTION: How does NASA organize a company party?
ANSWER: They planet.

But Martin wasn't satisfied with the standard formula. It bothered him that the laughter that followed a punchline was often automatic. Like Pavlov's dogs salivating at the sound of a bell, the audience would instinctively laugh when the punchline was delivered. What's more, if the punchline didn't produce laughter, the comedian would stand there, embarrassed, knowing his joke had bombed. Punchlines were a lousy way of doing comedy, Martin thought, both for the comedian and for the audience.

So Martin went back to first principles. He asked himself, *What if there were no punchlines? What if I created tension and never released it?* Instead of conforming to audience expectations, he decided to violate them. He believed that without a punchline, the resulting laughter would be stronger. The audience would laugh when they chose to do so, without being triggered by a gimmick.

Martin then did what all great rocket scientists do: He tested his idea. One night, he went onstage and told the audience that he was going to do the "Nose on Microphone" routine. He methodically placed his nose on the microphone, stepped back, and said, "Thank you very much."

There was no punchline. The audience sat in silence, stunned by Martin's departure from conventional comedy. But the laughter arrived when the audience caught up to what Martin had done. Martin's goal, in his words, was to leave the audience "unable to describe what it was that had made them laugh. In other words, like the helpless state of giddiness experienced by close friends tuned in to each other's sense of humor, you had to be there."

The initial response to Martin's first-principles approach was ridicule. One critic, sticking to the stand-up comedian's playbook, wrote, "This so-called 'comedian' should be told that jokes are supposed to have punch-lines." Another described Martin as "the most serious booking error in the history of Los Angeles."

That most serious booking error quickly became the most profitable one. Audiences and critics eventually caught up, and Martin became a stand-up legend.

But then he did something unimaginable. He quit.

Martin realized that he had achieved all that he could as a stand-up comic. If he had continued, his comedic innovations would have been minor deviations from the status quo. To save his art, he abandoned it.

Destruction, as the Red Hot Chili Peppers remind us in *Californication*, also breeds creation. Instead of withering, Martin's career blossomed. After he left stand-up, he acted in countless movies, recorded albums, and wrote books and screenplays. He won an Emmy, a Grammy, and an American Comedy award. At each stage, he learned, unlearned, and relearned.

I know firsthand how difficult it is to do what Martin did. When I first started my blog and podcast, venturing outside writing scholarly legal articles, a good friend and fellow law professor reached out to warn me. "You're ruining your scholarly significance," he said.

His comment reminded me of a line from a Dawna Markova poem: "I choose to risk my significance; to live so that which came to me as seed goes to the next as blossom."[24] When we look at the mirror, we tell ourselves a story. It's a story about who we are and who we aren't and what we should and shouldn't do.

We tell ourselves that we're a serious scholar, and serious scholars don't blog or podcast for the public. We tell ourselves that we're a serious comedian, and serious comedians don't leave their thriving stand-up career. We tell ourselves that we're a serious entrepreneur, and serious entrepreneurs don't pour their net worth into a risky space venture with little possibility for success.

There's a certainty to the story. The story makes us feel significant and secure. It makes us feel welcome. It connects us to the serious scholars, comedians, and entrepreneurs who came before us.

But instead of us shaping the story, it shapes us. Over time, the story becomes our identity. We don't change the story, because changing it would mean changing who we are. We fear losing everything we worked so hard to build, we fear that others might laugh, and we fear making fools of ourselves.

Like all others, the story of your significance is just that: a story. A narrative. A tale. If you don't like the story, you can change the story. Even better, you can drop it altogether and write a new one. "In order to change skins, evolve into new cycles," author Anaïs Nin writes, "one has to learn to discard."[25]

Discarding happened involuntarily for Steve Jobs, who in 1985 was forced out of Apple, the company he had cofounded. Although his dismissal stung at the time, looking back on it, Jobs says it was the "best thing that could have ever happened to me." Getting fired unshackled Jobs from his own history and forced him to return to first principles. "The heaviness of being successful was replaced by the lightness of being a beginner again. It freed me to enter one of the most creative periods of my life," Jobs says.[26] The baggage of his own perceived significance could no longer hold him back. His creative journey led him to start the computer company NeXT and join Pixar, turning the film company into a multi-billion-dollar success. He then returned to Apple in 1997 to launch a series of revolutionary products, such as the iPod and the iPhone.

It was agonizing for me to dismiss the advice of my well-intentioned friend who warned me against venturing into popular writing. There were moments of tremendous doubt along the way, when I thought I made the wrong call or perhaps should have stuck with my previous path. But if I had done that, you wouldn't be reading this book.

When we don't act—when we stick to the illusion of our significance—the risks are far greater. Only by leaving where we are can we get to where we want to go. You have to be "carbonized and mineralized," Henry Miller writes, "in order to work upwards from the last common denominator of the self."[27]

When you risk your significance, you won't change who you are. You'll discover it. As the ashes and clutter settle, something beautiful will soar.

One restaurant took this idea quite literally.

Appetite for Destruction

In 2005, chef Grant Achatz and his business partner Nick Kokonas founded the Chicago restaurant Alinea to create one of the world's best culinary experiences. "My hair was on fire to prove to the world what you can do with food," Achatz says.[28] Alinea's fire quickly illuminated the gastronomical world. Over thirty courses, the restaurant would delight its diners in an experience described as an "edible magic show" that continued to resonate in your mind and taste buds long after the meal ended.

Alinea achieved universal acclaim, collecting just about every award a restaurant could collect. In 2011, it became one of the first two restaurants in Chicago—and one of only nine in the United States—to earn the coveted three Michelin stars. The restaurant's tenth year in 2015 was its most profitable yet.

A celebration was in order. But this being Alinea, a conventional party wouldn't do. Kokonas had a different kind of party in mind—one involving sledgehammers.

In an interview, Kokonas recalled having a great meal at a prominent restaurant, only to return a few years later to great disappointment. "It's the same place, it's the same chair, it's the same meal more or less. Why is it so bad? Is it me? Did I change? Or is the world changing?" The answer, of course, was both.

"If you have a successful business," Kokonas explains, "it's actually harder to change it." The inertia required for changing course is too strong, particularly when you're at the top of your game. "It's hard to make incremental changes," he says. "Every now and then you just need to destroy it and rebuild it better."

Taking this statement to heart, Kokonas and his chef partner, Achatz, grew an appetite for destruction. They decided to jump off a creative cliff and gutted the restaurant from the inside out. Alinea closed down for five months, while both the building and the menu got a seven-figure transformation. The changes loosened "the sterile and hypercontrolled atmosphere that once made Alinea feel like the world's most pleasurable operating room," said one food critic.[29] The new restaurant offers the same gastronomical excellence but adds a good dose of fun and playfulness into the mix.

Foodies dubbed the new restaurant Alinea 2.0. But Kokonas and Achatz just call it Alinea. The restaurant may have been destroyed and rebuilt, but its core identity—and the founders' underlying commitment to first-principles thinking—remained unchanged.

This is an important point: Destruction, by itself, isn't enough if it's not accompanied by a commitment to the right thought process. "If a factory is torn down but the rationality which produced it is left standing, then that rationality will simply produce another factory," Robert Pirsig explains in *Zen and the Art of Motorcycle Maintenance*. "If a revolution destroys a systematic government, but the systematic patterns of thought that produced that government are left intact, then those patterns will repeat themselves."[30] Unless you change the underlying patterns of thought, you can expect more of the same—regardless of how many times you hold a sledgehammer party.

Changing the underlying patterns of thought requires hiring the right people. When interviewing prospective team members, Kokonas doesn't "want people that have 20 years of experience working in restaurants." Too much baggage can get in the way of first-principles thinking. Seasoned employees, Kokonas worries, will look at a restaurant and think white tablecloths.

If you're trying to transform an industry, it makes sense to look outside the industry for talent. That's where you'll find people who aren't blinded by the invisible rules—the white tablecloths—that constrain their thinking. In its early days, SpaceX would often hire people from the automotive and cell-phone industries. These are fields where technologies change rapidly, requiring quick learning and adaptation—the hallmark of first-principles thinkers.

....

WHAT'S REMARKABLE ABOUT both Steve Martin and Alinea is that they took a sledgehammer to themselves when they were at the top of their game. But most of us can't stomach doing what Martin and Alinea did. When things are going well, we settle into the comfort of the status quo, rather than upending it.

But returning to first principles is easier than you might assume. If you can't use an actual wrecking ball, you can try a hypothetical one.

I Came in Like a Wrecking Ball

Kenneth Frazier's story is quintessentially American. The son of a janitor, Frazier grew up in a working-class neighborhood in Philadelphia and climbed to the top, graduating from Penn State and then Harvard Law School. He joined the pharmaceutical giant Merck as corporate counsel and eventually became its CEO.[31]

Like most executives, Frazier wanted to promote innovation at Merck. But unlike most executives who simply ask their employees to innovate, Frazier asked them to do something they had never done before: destroy Merck. Frazier had the company executives play the role of Merck's top competitors and generate ideas to put Merck out of business. They then reversed their roles, went back to being Merck employees, and devised strategies to avert these threats.[32]

This is called the kill-the-company exercise. As Lisa Bodell, the mastermind behind the exercise, explains, "To create the company of *tomorrow*, you must break down the bad habits, silos, and inhibitors that exist *today*."[33] These habits are difficult to break down because we often adopt the same internal perspective. It's like trying to "psychoanalyze yourself,"

Bodell says. We're too close to our own problems and weaknesses to evaluate them objectively.

The kill-the-company exercise forces you to switch perspectives and play the role of an antagonist who doesn't care about your rules, habits, and processes. The participants must employ first-principles thinking, use new neural pathways, and come up with original ideas that move beyond mere platitudes. It's one thing to say "let's think outside the box." It's another to actually step outside the box and examine your company or product from the viewpoint of a competitor seeking to destroy it. By viewing our weaknesses through this out-of-company experience, we realize we may be standing on a burning platform. The urgency of change becomes clear.

The US military also uses a version of the kill-the-company exercise in war-gaming. It's called *red teaming*, a term that's a relic of the Cold War. In simulations, the red team plays the role of the enemy and finds ways to scuttle the blue team's mission. Red teams expose the flaws in planning and execution so that the problems can be fixed before the mission begins. As Major Patrick Lieneweg, who teaches red-teaming seminars, explained to me, the process plays a critical role in mitigating groupthink in the otherwise hierarchical environment of the military: "It improves the quality of thinking by challenging prevailing notions, testing assumptions, and asking critical questions."

Bezos follows a similar approach at Amazon.[34] When ebooks began to threaten Amazon's physical book business, Bezos embraced the challenge instead of turning away from it. He told one of his associates, "I want you to proceed as if your job is to put everyone selling books out of a job," including Amazon itself. The business model this exercise produced eventually shot Amazon to the top of the ebook market.

I also applied a version of the kill-the-company exercise in my law-school classroom. In my classes on authoritarian regimes, I would lecture my students on how modern dictators have abandoned the openly repressive tactics of their predecessors. Today's authoritarians frequently come to power by democratic elections and then erode democracy through seemingly legitimate means. They conceal authoritarian tactics under the trappings of democracy.

Although I would warn my students that no country—including the United States—is immune to these stealth authoritarian threats, I sensed that these lectures never really resonated. My students assumed that authoritarian takeovers happen only in backward, faraway lands, in countries riddled with corruption and incompetence, and in nations that end with -*stan*.

So I decided to go rogue.

I threw away my lecture notes and instead asked my students to conduct a thought experiment: Play the role of an aspiring dictator and come up with ways to decimate democracy in the United States. They then switched roles and devised measures to guard against the most serious threats.

Here's the thing: When we talk in the abstract about protecting American democracy, the urgency to do so isn't clear. After all, the democratic system in the United States has shown tremendous resilience. But when we put ourselves in the shoes of a dictator and actually devise strategies to decimate American democracy, the weak points in the system reveal themselves. Only when we realize the fragility of the system do we recognize the imperative to protect it.

The kill-the-company exercise isn't just for megacorporations or law-school classrooms. You can employ variations of it in your own life by asking questions like the following:

- Why might my boss pass me up for a promotion?
- Why is this prospective employer justified in not hiring me?
- Why are customers making the right decision by buying from our competitors?

Avoid answering these questions as you would that dreadful interview prompt, "Tell me about your weaknesses," which tends to induce humble-bragging ("I work too hard"). Instead, really get into the shoes of the people who might reject your promotion, refuse to hire you, or buy from your competitors. Ask yourself, *Why are they making that choice?*

It's not because they're stupid. It's not because they're wrong and you're right. It's because they see something that you're missing. It's because they believe something you don't believe. And you can't change that worldview or that belief by calling the same plays from the same old playbook. Once you've got a good answer to these questions, switch perspectives and find ways to defend against these potential threats.

But you don't always need an actual or a hypothetical wrecking ball to get back to first principles. Sometimes a razor will do.

Occam's Razor

Legend has it that NASA spent a decade and millions of dollars developing a ballpoint pen that would work in zero gravity and function in extreme temperatures. The Soviets used a pencil.

The story of the "write stuff" is a myth.[35] Pencil tips have a habit of breaking and getting into nooks and crannies—which may be okay on Earth, but not okay on a spacecraft, where they can find their way into mission-critical equipment or end up floating into an astronaut's eyeball.

But the moral of the myth still holds. As Einstein said, everything should be made "as simple and as few as possible."[36] This principle is known as Occam's razor. The name, I admit, is unfortunate. It sounds like a cheap late-night horror flick, but it's actually a mental model named after William of Ockham, a fourteenth-century philosopher. The model is often stated as a rule: The simplest solution to a problem is the correct one.

This popular description happens to be wrong. Occam's razor is a guiding principle—not a hard-and-fast rule. Nor is it a preference for the simple at all costs. Rather, it's a preference for the simple, *all other things being equal.* Carl Sagan put it well: "When faced with two hypotheses that explain the data *equally well,*" you should "choose the simpler."[37] In other words, "when you hear hoofbeats, think horses, not unicorns."[38]

71

Occam's razor cuts through the clutter that often gets in the way of first-principles thinking. The most elegant theories rest on the fewest assumptions. The most elegant solutions, the rocket scientist David Murray writes, "use the least number of components to solve the greatest number of problems."[39]

Simple is sophisticated. Newton's laws of motion, for example, are poetic in their simplicity. Take his third law: For every action, there is an equal and opposite reaction. Centuries before the advent of human flight, this simple law explained how rockets reach space. The mass of their fuel goes down, and the rocket goes up.

"The more we understand something," Peter Attia explained to me, "the less complicated it becomes. This is classic Richard Feynman teaching." Attia is a mechanical engineer turned medical doctor, a renowned expert in increasing people's life span and health span. If you're reading a study in medicine, he said, "and you see words like *multifaceted, multifactorial, complex*, to explain the current understanding," the authors are basically saying, "We don't know what the heck we're talking about yet." But when we really understand the cause of a disease or an epidemic, "it tends to be simple and not multifactorial."[40]

Simple also has fewer points of failure. Complicated things break more easily. This principle is as true in rocket science as it is in business, computer programming, and relationships. Every time you introduce complexity to a system, you're giving it one more aspect that can fail. As the safety manager for Apollo 8 remarked, the spacecraft had 5.6 million parts, and "even if all functioned with 99.9 percent reliability, we could expect 5,600 defects."[41]

Simplicity also reduces costs. The Atlas V rocket—which has taken many objects, including military satellites and Mars rovers, into space—uses up to three types of engines for different stages of flight.[42] This complexity drives up the expenses: "To a first-order approximation," Musk explains, "you've just tripled your factory costs and all your operational costs."

In contrast, SpaceX's Falcon 9 has two stages with the same diameter and the same engines built from the same aluminum-lithium alloy. This simplicity enables high-volume production at a lower cost, while boosting reliability. What's more, unlike other aerospace companies that build their vehicles vertically—in the same position that they're launched—SpaceX assembles them horizontally.[43] This orientation allows the company to use a regular warehouse, eliminating the need to build a skyscraper—not to mention the safety issues that come with workers dangling sixty feet in the air while building a rocket. "Every decision we've made," Musk says, "has been with consideration to simplicity. . . . If you've got fewer components, that's fewer components to go wrong and fewer components to buy."[44]

The Russians adopted a similar approach for the Soyuz launcher, used for transporting crew and cargo to the International Space Station. Soyuz is considered more reliable than NASA's space shuttle in part because it's a "much simpler vehicle to operate," astronaut Chris Hadfield writes.[45] Paolo Nespoli, another astronaut, put it this way: "We could learn a lot from the Russians that sometimes when you do less, less is better."[46]

The noise in any system—whether it's a rocket, a business, or your résumé—reduces its value. There's a temptation to always add more, but the taller the Jenga tower, the more fragile it gets. "Any intelligent fool can make things bigger and more complex," economist E. F. Schumacher said in a quote often misattributed to Einstein. "It takes a touch of genius and a lot of courage to move in the opposite direction."[47]

In rocket science, Natalya Bailey, the thirty-three-year-old founder and CEO of aerospace start-up Accion, is at the forefront of this movement in the opposite direction. As a child, she would lie on a trampoline outside her family home in Newberg, Oregon, and gaze at the stars. Among the usual collection of twinkling stars, Bailey once spotted solid lights moving steadily across the sky. These, she later learned, were spent stages of rockets. "That blew my mind," Bailey told me.

The trampoline stargazer eventually decided to get a college degree in aerospace engineering and a PhD in space propulsion. During her education, she became interested in rockets that use electric energy to propel themselves. "All rockets work by the same principle. Throwing stuff out the back pushes the spacecraft forward," Bailey told me, referring to Newton's third law of motion. For traditional, chemical rockets, that stuff is hot gases. But for an electric engine, that stuff is ions—molecules with electric charges.

Chemical rockets work well for getting a spacecraft into orbit because they can produce a lot of thrust very quickly. Electric propulsion, in contrast, is much slower, but it's ten to a hundred times more energy-efficient. Electricity is also safer to use since it doesn't require toxic propellants or pressurized tanks.[48] As part of her PhD dissertation, Bailey started designing tiny electric-propulsion engines. That research became the basis for the aerospace company she cofounded, Accion—named after a summoning charm from the Harry Potter books.

Accion's engines are fired after a satellite has been placed in orbit. The size of a deck of cards, the engine can push satellites as big as refrigerators and move them around while they're floating in orbit. Equipped with these engines, satellites can linger in orbit longer and avoid colliding with the nearly eighteen thousand other pieces of human-made debris and junk circling the planet.[49] The technology also has the potential to help propel spacecraft to other planets. With Accion's technology, you can use a shoebox-sized engine and fuel system, instead of giant fuel tanks, to take a spacecraft to Mars once it's in Earth orbit.[50]

Bailey is just like her engines: She's humble and understated, but packs tremendous punch. What SpaceX and Blue Origin are doing for rockets, Bailey and her Accion team are doing for the satellites those rockets carry into space.

As these examples show, simple can be mighty. But don't confuse simple with easy. As the quote attributed to many luminaries says, "If I had more time, I would have written a shorter letter." We admire the simplicity of Newton's laws and Accion's engines, but we don't see the messy and complex precursors that these scientists had to winnow down through tremendous effort.[51]

Physics has a way of forcing rocket scientists to use Occam's razor. Weight and space are at a premium in spacecraft design. The heavier the spacecraft is, the more expensive it is to design and launch. Rocket scientists have to constantly ask themselves, *How can we possibly make this fit into that?* They get the right fit by cutting out the junk, reducing the system to its irreducible minimums, and making everything as simple as possible without compromising the mission.

If you want to soar, you must cut what weighs you down. You can take your cue, once again, from Alinea. Achatz explains that when he and Kokonas opened the restaurant, "one of our creative roads was to look at a dish on paper or in front of us and ask, 'What else? What else can we do? What else can we add? What can we add to make this better?'"[52] But over time, they reversed their approach. "Now," Achatz says, "we find ourselves constantly asking, 'What can we take away?'" Michelangelo approached sculpting in the same way. As he explained, "The sculptor arrives at his end by taking away what is superfluous."[53]

Paint yourself a vivid picture of the future with the excesses wiped off your plate. What does it look like? Ask yourself, as one innovative CEO does, "What if you had not already hired this person, installed this equipment, implemented this process, bought this business, or pursued this strategy? Would you do the same thing you are doing today?"[54]

Like all sharp objects, Occam's razor can cut both ways. In some cases, the complex solution will lead to a better result. Don't use Occam's razor to validate the natural human craving for simplicity in the face of nuance and complexity. Don't confuse a simple solution, as H. L. Mencken cautioned, with one that is "neat, plausible, and wrong."[55] Even as you seek to simplify, remain open to new facts that complicate matters. As the English mathematician and philosopher Alfred North Whitehead put it, "Seek simplicity and distrust it."[56]

To cut is to make whole. To subtract is to add. To constrain is to liberate.

The virtues of cutting—of returning to the origin to find the original—should remind you that what you need isn't out there waiting to be discovered in a competitor's playbook or a role model's life story. It's already here.

Once you've returned to first principles—cut the assumptions and processes cluttering your thinking—it's time to unleash the most complex and innovative instrument available at your disposal: your own mind.

Visit **ozanvarol.com/rocket** to find worksheets, challenges, and exercises to help you implement the strategies discussed in this chapter.

3

A MIND AT PLAY

How to Ignite Breakthroughs
with Thought Experiments

When I examine myself and my methods of thought,
I come to the conclusion that the gift of fantasy has meant more
to me than my talent for absorbing positive knowledge.
—ALBERT EINSTEIN

W HAT WOULD HAPPEN if I chased after a beam of light?[1] A sixteen-year-old Albert Einstein pondered this question after he had run away from his unimaginative German school that emphasized rote memorization at the expense of creative thinking. His destination was a reformist Swiss school founded on the principles of Johann Heinrich Pestalozzi, who championed learning through visualization.

While there, Einstein put Pestalozzi's principles into action and visualized himself chasing after a beam of light. He believed that if he managed to catch up with it, he would observe a frozen light beam. This belief, which conflicted with Maxwell's equations on the oscillations of electromagnetic fields, caused Einstein what he described as "psychic tension." The resolution of this psychic tension took him a decade—and eventually produced the special theory of relativity.

It was another question that later produced the general theory of relativity: Does a person who falls freely in an enclosed chamber feel his own weight?

This question—which Einstein later called "the happiest thought of my life"—occurred to him while he was daydreaming at his desk in a Swiss patent office. Einstein's job as a patent clerk had trained him well for visualization. Evaluating patent applications required him to picture how each invention would work in practice. Picturing his new thought experiment, he concluded that the man falling freely would *not* feel his own weight and would instead think he was floating in zero gravity. This conclusion, in turn, led him to another major discovery: Gravity and acceleration are the same.

Einstein credits these thought experiments (or what he would have called *Gedankenexperimente*) for virtually all his breakthroughs. Over his lifetime, he would visualize "lightning strikes and moving trains, accelerating elevators and falling painters, two-dimensional blind beetles crawling on curved branches," among many others.[2] With his mind at play, Einstein upended entrenched assumptions in physics, cementing himself as one of the most popular scientific figures in the public imagination.

This chapter is about the power of thought experiments. You'll discover why the key to supercharging your creativity is to do nothing at all and how most work environments sabotage, rather than boost, people's creative potential. You'll learn why you should compare apples and oranges and what made Newton the least favorite professor on campus. I'll reveal how a simple question from an eight-year-old created a billion-dollar author and what a revolutionary running shoe and one of the greatest rock songs of all time have in common. Along the way, you'll meet scientists, musicians, and entrepreneurs who have used a technique called combinatory play to produce breakthrough works, and you'll learn how to apply it in your own life.

The Laboratory of the Mind

Although they're associated with Einstein in popular culture, thought experiments date back at least to the ancient Greeks. Since then, they have spread across disciplines, generating breakthroughs in philosophy, physics, biology, economics, and beyond. They have powered rockets, toppled governments, developed evolutionary biology, unlocked mysteries of the cosmos, and created innovative businesses.

Thought experiments construct a parallel universe in which things work differently. They require us, as philosopher Kendall Walton explains, to "imagine specific fictional worlds, as kinds of situational setups that when you run, perform, or simply imagine them, lead to specific results."[3] Through thought experiments, we transcend everyday thinking and evolve from passive observers to active interveners in our reality.[4]

If the brain had a tail, thought experiments would make it wag.

There are no precise spells for conducting thought experiments or secret recipes you can copy. Formulas and rules are antithetical to first-principles thinking, so every well-crafted thought experiment is unique in its own way. In this chapter, I'll help you create the right conditions for thought experiments, but my intention is to guide, not to constrain.

When we think of scientists, we often imagine lab-coated brainiacs poring over state-of-the-art microscopes in fluorescent-lit labs. But for many scientists, the laboratory of the mind is far more important than the laboratory of the physical world. As rockets fire spacecraft, thought experiments fire our neurons.

Consider Nikola Tesla, the famous Serbian American inventor. His thought experiments powered his imagination, producing the alternating-current system that now powers our lives.[5] Tesla built and tested inventions all in his mind. "Before I put a sketch on paper, the whole idea is worked out mentally," he explained. "I do not rush into actual work. When I get an idea, I start at once building it up in my imagination. I change the construction, make improvements and operate the device in my mind. It is absolutely immaterial to me whether I run my turbine in thought or test it in my shop."[6]

Leonardo da Vinci did the same. He famously used his notebooks for thought experiments, sketching various engineering designs he formulated in his mind—from flying machines to churches—instead of physically constructing them.[7]

Let's pause there for a moment. As shocking as it sounds, we can generate breakthroughs simply by thinking. No Google. No self-help books. No focus groups or surveys. No advice from a self-proclaimed life coach or an expensive consultant. No copying from competitors. This external search for answers impedes first-principles thinking by focusing our attention on how things *are* rather than how they *could* be.

Thought experiments take this external inquiry and turn it inward—just you and your imagination. "Pure thought," Einstein said, "can grasp reality."[8] Thoughts can disprove an argument, show why something will or won't work, and illuminate the way forward—all without a single physical experiment.

Consider this example. In a world with no air resistance, if you simultaneously drop a heavy bowling ball and a light basketball from the same height, which one would strike the ground first? Aristotle believed that the heavy object would fall faster than the lighter one. This theory persisted for two millennia, until an Italian scientist named Galileo Galilei entered the scene. Galileo was born a misfit in a world of conformists. He challenged tyrannical dogma across diverse disciplines, most famously championing heliocentrism, which placed the Sun, rather than the Earth, at the center of the solar system.

Galileo also took on Aristotle's theory. The Italian didn't believe that acceleration increased with mass. So he climbed to the top of the Leaning Tower of Pisa, dropped two objects of different weights, and chuckled in delight while calling Aristotle funny names when both objects hit the ground at the same time.

Except he didn't.

This entire episode turned out to be a myth manufactured by Galileo's earliest biographer. Most contemporary historians agree that Galileo instead conducted a thought experiment—not a physical one. He imagined a heavy cannonball and a light musket ball chained together to form a single, combined system to be dropped at the same time.[9] If Aristotle were right, the attached system should fall faster than the heavy cannonball alone because the combination weighs more. But it would also mean the light musket ball in the attached system should fall slower than the heavy cannonball. In other words, if Aristotle's theory is correct, the light musket ball should act as a drag on the combined system, causing it to fall slower than the heavy cannonball alone.

Both statements can't be true: The attached system can't fall both faster and slower than the heavy cannonball on its own. The thought experiment reveals a contradiction in Aristotle's theory and obliterates it. Through thought alone and without spending a dime, a venerated theory was cast aside, making room for a new one.

Centuries later, Galileo's thought experiment was put to the test on the Moon. In 1971, during the Apollo 15 mission, astronaut David Scott dropped a hammer and a feather from the same height while standing on the Moon's surface. Both fell at the same rate and struck the lunar surface at the same time. The official science report described the result as "reassuring" in light of "both the number of viewers that witnessed the experiment and the fact that the homeward journey was based critically on the validity of the particular theory being tested."[10]

Curiosity is a crucial ingredient in any thought experiment. It's what propelled Galileo to pose his thought experiment, and Scott to test its validity on the lunar surface. Yet for much of society, curiosity isn't a great virtue, but is a killer vice.

Curiosity Killed Schrödinger's Cat

Can a cat be alive and dead at the same time? This was the question that Austrian physicist Erwin Schrödinger asked through a famous thought experiment.[11] His goal was to stretch the limits of what's known as the Copenhagen interpretation of quantum mechanics. According to the interpretation, quantum particles (such as atoms) exist in a combination—or superposition—of different states. Put differently, a quantum particle can be in two states or in two places simultaneously. It's only when someone observes the particle that it collapses into one of many possible states.

Schrödinger took the Copenhagen interpretation and applied it to a cat. In his thought experiment, a cat is placed inside a sealed box with a vial of poison that will be randomly released when a radioactive substance inside the box decays. If you buy the Copenhagen interpretation, before the box is opened, the cat is in a superposition: It's both alive and dead. Only when someone opens the box does the cat collapse into one of these two realities.

This result, of course, is wildly counterintuitive. But it was precisely the point of Schrödinger's thought experiment—to contradict, to provoke, and to disprove the Copenhagen interpretation by taking it to its logical extremes.

But there's another takeaway from this thought experiment. It wasn't the poison that killed the cat. It was the act of curious observation, of not minding your own business, of opening the box to see what's inside the same way a child might sneak to open presents the night before Christmas.

There's an idiom in the English language dedicated to this idea: Curiosity killed the cat. Or as Russians say with far more dramatic flair, "The nose of curious Barbara was torn off at the market."[12]

These idioms, according to the ever-reliable Wikipedia, are "used to warn of the dangers of unnecessary investigation or experimentation." Curiosity, in cats or in Russian market-goers, isn't just annoying or inconvenient. People who ask questions or pose thought experiments aren't just pesky troublemakers who can't be satisfied with the status quo. They're downright dangerous. As the renowned Hollywood producer Brian Grazer and his coauthor Charles Fishman write, "The child who feels free to ask why the sky is blue grows into the adult who asks more disruptive questions: Why am I the serf and you the king? Does the sun really revolve around Earth? Why are people with dark skin slaves and people with light skin their masters?"[13]

We discourage curiosity also because it requires an admission of ignorance. Asking a question or posing a thought experiment means that we don't know the answer, and that's an admission that few of us are willing to make. For fear of sounding stupid, we assume most questions are too basic to ask, so we don't ask them.

What's more, in this era of "move fast and break things," curiosity can seem like an unnecessary luxury. With an inbox-zero ethos and an unyielding focus on hustle and execution, answers appear efficient. They illuminate the path forward and give us that life hack so we can move on to the next thing on our to-do list. Questions, on the other hand, are exceedingly inefficient. If they don't yield immediate answers, they're unlikely to get a slot on our overloaded calendars.

At best, we pay lip service to curiosity but end up discouraging it in practice. Businesses hold a "creativity day" to foster innovation—complete with a PowerPoint presentation and an expensive outside speaker—but go back to business as usual for the remaining 364 days. Employees are rewarded for staying the course rather than questioning it. According to a survey of workers in sixteen industries, "while 65% said that curiosity was essential to discover new ideas, virtually the same percentage felt unable to ask questions on the job."[14] Although 84 percent in the same survey said their employers encouraged curiosity on paper, 60 percent encountered barriers to it in practice.

Instead of making curiosity the norm, we wait until a crisis occurs to become curious. Only when we're laid off do we begin to ponder alternative career paths. And only when our business is disrupted by a young, scrappy, and hungry competitor do we gather the troops to spend a few futile hours to "think outside the box."

For answers, we rely on the same methods, the same brainstorming approaches, and the same stale neural pathways. It's no wonder that the resulting innovations aren't innovations at all. They're at best insignificant deviations from the status quo. Look at any monstrous company or bloated bureaucracy collapsing under its own weight, and you'll find a historical lack of curiosity.

Fear of the outcome is another reason we shun curiosity. We don't ask hard questions when we're afraid of what we might find (which is why people are reluctant to visit their doctor when they fear the diagnosis). Worse, we're afraid that we may not find anything at all—that our inquiry led us nowhere—turning this whole thought-experiment business into a gigantic waste of time.

We also assume that thought experiments require complex mental gymnastics or divine inspiration. We tell ourselves that someone far smarter than us would have already posed the question if it were worth asking.

But geniuses don't have a monopoly on thought experiments. There are no chosen few. You don't need Einstein's electrified head of hair to conduct thought experiments. It may not feel like it, but we're all experimenters at heart—walking repositories of epiphanies hidden in our subconscious.

Seemingly unnecessary investigation and experimentation are precisely what you need to uncover those epiphanies. George Bernard Shaw once said, "Few people think more than two or three times a year. I've made an international reputation for myself by thinking once or twice a week."[15] As Shaw knew, hustle and creativity are antithetical to each other. You can't generate breakthroughs while clearing out your inbox. You must dig the well before you're thirsty and become curious *now*—not when a crisis inevitably presents itself.

Curiosity may have killed Schrödinger's cat. But it just might save you.

A Lifelong Kindergarten

"Why can't I see the picture now?"[16] It was 1943, and Edwin Land was vacationing with his family in Santa Fe, New Mexico. Land, the cofounder of Polaroid and a camera enthusiast, was taking photos of his three-year-old daughter, Jennifer. Back then, there was no instant photography. The film had to be developed and processed in a darkroom before the photos could see the light of day—a turnaround time of several days. Although there are conflicting reports of what exactly happened, according to one popular account, a question that the precocious Jennifer asked her father changed everything.

"Why can't I see the picture now?" Land took this question to heart. But he faced a big constraint. A huge darkroom couldn't fit inside a small camera. He set out for a long walk to ponder the problem and came up with a thought experiment. What if the camera carried a small reservoir containing the chemicals used to develop film in a darkroom? The chemicals would be spread over a negative film and released onto the positive layer, producing the final image.

It took several years to perfect the technology, but the thought experiment eventually led to the invention of instant photography. With the new technology, only seconds, not days, would pass between the click of the shutter and a physical photo in your hands.

Although thought experiments don't come naturally to most adults, we mastered them as children. Before the world stuffed us with facts, memos, and right answers, we were moved by genuine curiosity. We stared at the world, wrapped in awe, and took nothing for granted. We were blissfully unaware of social rules and viewed the world as our very own thought experiment. We approached life not with the assumption that we knew (or should know) the answers, but with the desire to learn, experiment, and absorb.

A favorite example of mine is about a kindergarten teacher who was walking around the room to check each child's work as they drew pictures. "What are you drawing?" he asked one student. The girl said, "I'm drawing God." The teacher was shocked at this deviation from the standard curriculum. He said, "But no one knows what God looks like." The girl replied, "They will in a minute."

Children intuitively grasp one cosmic truth that eludes most adults: It's all just a game—a big, marvelous game. In the popular children's book *Harold and the Purple Crayon*, the four-year-old protagonist has the power to create things just by drawing them. There's no path to walk on, so he draws a path. There's no moon to light his path, so he draws the moon. There are no trees to climb on, so he draws an apple tree. Throughout the story, his imagination brings things into existence.[17]

Thought experiments are your very own reality-distortion field, your choose-your-own-adventure game—your purple crayon.

The purple crayon was Einstein's favorite scientific tool, one that he carried with him even as an adult.[18] As he wrote to a friend, "You and I never cease to stand like curious children before the great mystery into which we were born."[19] Centuries earlier, Isaac Newton purportedly used similar words in describing himself as "a boy playing on the seashore . . . whilst the great ocean of truth lay all undiscovered before me."[20]

Although Einstein and Newton managed to retain their childlike curiosity, it is beaten out of most people. Our conformist education system, designed to churn out industrial workers, is partly to blame ("No one knows what God looks like"). Our natural curiosity is also suppressed by busy, well-meaning parents who believe that everything important has already been settled. One can imagine an annoyed father in Edwin Land's place dismissing his daughter's question as absurd ("Patience, Jennifer! Learn to wait for the photo"). Or a busy mother missing the genius in sixteen-year-old Einstein's beam-riding experiment ("Go back to your room, Albert. And stop the crazy talk").

Over time, as we settle into adulthood, as student loans and mortgages begin to mount, our curiosity is replaced by complacency. We view intelligent urges as a virtue and playful urges as a vice.

But play and intelligence should be complements, not competitors. Play, put differently, can be a portal to intelligence. In his seminal article "The Technology of Foolishness," James March writes that "playfulness is a deliberate, temporary relaxation of rules in order to explore the possibilities of alternative rules."[21] He argues that individuals and organizations "need ways of doing things for which they have no good reason. Not always. Not usually. But sometimes." Only by taking a playful attitude toward our own beliefs can we challenge and change them.

The operative word in a thought experiment is *experiment*. This framing should lower the stakes. A thought experiment sets up a sandbox in the controlled environment of your mind. If it doesn't work, nothing bad happens. There's no collateral damage or spillover effects. At the initial stage, you're not aiming for implementation—let alone perfection—so you're less likely to get hamstrung by your assumptions, biases, and fears.

Recapturing our childlike curiosity can boost originality—and there's plenty of research to back this up.[22] Yet being told to think like a child can feel like being ordered to stay dry in a thunderstorm.

Here's the good news: You can capture a childlike curiosity without physically regressing to your childhood or developing Peter Pan syndrome. Reconnecting with your inner child might be as easy as pretending to be a seven-year-old. This suggestion sounds bizarre, but it works. In one study, when participants were instructed to imagine themselves as seven-year-olds with free time, they performed better in objective tests of creative thinking.[23] For this reason, the MIT Media Lab—devoted to "the unconventional mixing and matching of seemingly disparate research areas"— has a section called Lifelong Kindergarten.[24]

Minds are far more malleable than we assume. If we pretend that life is one long kindergarten, our minds just might follow.

....

AT THIS STAGE, you might be wondering, if the thought experiment doesn't make sense, if it more properly belongs in a children's game, what's the point? If the thought experiment can't be implemented, what—if anything—separates it from useless fantasy?

The purpose of a thought experiment isn't to find the "right answer"— at least not initially. This isn't like your high-school chemistry class, where the outcome of each experiment was predetermined, leaving no room for curiosity or unexpected insights. If you didn't get the right result, you would be stuck in the lab fiddling with test tubes and beakers, while your classmates trekked off to the movies. The point of Einstein's thought experiment wasn't to figure out a way to actually ride next to a light beam. Rather, it was to ignite a process of open-minded inquiry that can—and often did—result in unexpected major insights.

Pursuing a thought experiment—even one that leads nowhere—can lead to breakthroughs. Fantasies, as Walter Isaacson writes, can be "paths to reality."[25] It's a little like driving from New York to Hawaii. Impossible? Yes. Will you discover profound new insights along the way before you hit the giant practical constraint that's the Pacific Ocean? Absolutely. The goal is to jolt you out of your autopilot mode, keeping your mind receptive to possibilities.

Remember, the thought experiment is the starting point, not the end. The process is messy and nonlinear. And the answer, as we'll see in the next section, will often come when you're least expecting it.

Get Bored More Often

I couldn't remember the last time I was bored.

I had just woken up and grabbed my phone to take my morning dose of digital notifications. As I was about to start scrolling through my various feeds, I had an epiphany.

I couldn't remember the last time I was bored.

Along with my VHS player and Bon Jovi cassettes, boredom had become a relic of the past. Gone were the days when I would lie awake in bed in the morning, bored out of my mind, and daydream for a while before deciding to immerse myself into reality. I would no longer twiddle my thumbs while waiting for a haircut or strike up a conversation with a stranger waiting in line at a coffee shop.

I viewed boredom—which I define as large chunks of unstructured time free of distractions—as something to be avoided. Boredom evoked memories of getting chastised by teachers for daydreaming. Boredom, for me, was a bitter cocktail of agitation, impatience, and despair. I assumed that only boring people got bored, so I filled—no stuffed—every moment of my day with activity.

I know I'm not alone here. On a typical day, we switch from one form of social media to the next, check our email, catch up on the news—all within a span of twenty minutes. We prefer the certainty of these distractions over the uncertainty of boredom (*I don't know what to do with myself, and I'd rather not find out*). In a 2017 survey, roughly 80 percent of Americans reported that they spent no time whatsoever "relaxing or thinking."[26]

During rare moments of tranquility, we feel almost guilty. As notifications scream their hundred-decibel sirens for attention, we feel compelled to take a furtive glance at them so we don't miss out. Rather than being proactive, we spend most of our days—and our lives—playing defense. We self-soothe with the same distractions that ultimately make us feel worse.

Our responses stoke the fire, rather than put it out. Each email we send generates even more emails. Each Facebook message and tweet gives us a reason to return. It's a Sisyphean torture, endlessly rolling a boulder up an impossible hill.

Yet we prefer this torture over boredom. In a 2014 study, researchers placed college-age participants in a room, removing all their belongings.[27] They left the participants to their own devices and told them to spend time with their thoughts for fifteen minutes. I know, fifteen minutes—*yikes!*—but that's why the researchers gave the internet-bred participants a choice: If they wanted, instead of losing themselves in thought, the students could self-administer an electric shock by pressing a button. In the study, 67 percent of men and 25 percent of women chose to shock themselves instead of sitting undisturbed with their thoughts (including one person who delivered a whopping 190 shocks to himself during the fifteen-minute period).

A shocking thought indeed.

Boredom, in other words, is now an endangered state. This isn't an innocuous development. Without boredom, our creativity muscles begin to atrophy from disuse. "We are drowning in information," biologist E. O. Wilson said, "while starving for wisdom."[28] If we don't take the time to think—if we don't pause, understand, and deliberate—we can't find wisdom or form new ideas. We end up sticking with the first solution or thought that pops into our mind, instead of staying with the problem. But problems worth solving don't yield immediate answers. As author William Deresiewicz explains, "My first thought is never my best thought. My first thought is always someone else's; it's always what I've already heard about the subject, always the conventional wisdom."[29]

We appear to be deferring life when we get bored, but it's quite the opposite. In one study, two British researchers culled through decades of research, concluding that boredom should "be recognized as a legitimate human emotion that can be central to learning and creativity."[30] Falling into boredom allows our brain to tune out the external world and tune into the internal. This state of mind lets loose the most complex instrument known to us, switching the brain from the focused to the diffused mode of thinking. As the mind begins to wander and daydream, the default mode network in our brain—which, according to some studies, plays a key role in creativity—lights up.[31]

As the saying goes, it's the silence between the notes that makes the music.

Isaac Newton was "the least popular professor" on campus because "he'd stop in the middle of a lecture with a creative pause that could extend for minutes," while his students waited for him to return to earth.[32] During that pause, nothing appeared to be happening, but appearances deceive. Even when it's idling, the brain is still active.[33] "When you're staring into space," Alex Soojung-Kim Pang writes, "your brain consumes only slightly less energy than it does when you're solving differential equations."[34]

So where does all that energy go? Your mind may seem to be drifting from one irrelevant topic to the next, but your subconscious is hard at work, consolidating memories, making associations, and marrying the new with the old to create novel combinations.[35] The phrase *unconscious mind* is an insult to a part of our brain that does so much work behind the scenes.

When we sit still, we turn into a magnetized rod that attracts ideas. This is why phrases like *epiphany, flash of light,* or *stroke of genius* are often used to describe the eureka moment—Greek for "I've found it." Ideas seem to explode into life during slack times, not during hard labor. Einstein was daydreaming when he had the revelation—a person who falls freely doesn't feel his own weight—that led to the general theory of relativity. Danish physicist Niels Bohr literally dreamed up the structure of an atom when he envisioned himself "sitting on the sun with all the planets hissing around on tiny cords."[36] Archimedes's famous eureka moment purportedly arrived when he was easing himself into a bath.[37]

There's a TV commercial where business executives squeeze themselves into a shower at work. One person asks, "Why are we meeting in the shower?" The boss replies, "Well, ideas always hit me in my shower at home."[38]

The idea-in-the-shower moment is cliché because it works. The fix for a defective mirror on the Hubble Space Telescope was dreamed up in the shower. Launched in 1990 to take high-resolution images of space, the telescope suffered from fuzzy vision because of a defective mirror. The fix required astronauts to reach deep into the telescope's belly—not an easy feat for a satellite barreling around the Earth several hundred miles above its surface. While staying in a German hotel room, NASA engineer James Crocker stumbled on an adjustable shower head that extended or retracted to suit different heights. This observation was Crocker's aha moment. He devised a solution to do the same for the Hubble by using automated arms that extended to reach into the seemingly inaccessible parts of the telescope.[39]

These epiphanies appear effortless, but they're the product of a long, slow burn. A breakthrough begins with asking a good question, laboring over the answer intensely, and being stuck in idleness for days, weeks, and sometimes years. Research shows that incubation periods—the time you spend feeling stuck—boosts the ability to solve problems.[40]

As we saw earlier, Andrew Wiles became a mathematical celebrity after proving Fermat's last theorem. Being stuck, according to Wiles, is "part of the process."[41] But "people don't get used to that," he says. "They find it very stressful." When he got stuck—which was often—Wiles would stop, let his mind relax, and go for a walk by the lake. "Walking," he explains, "has a very good effect in that you're in this state of relaxation, but at the same time you're allowing the sub-conscious to work on you."[42] As Wiles knew, a watched pot never boils. You often have to walk away from the problem—literally and metaphorically—for the answer to arrive.[43]

A good footslogging is part of many scientists' tool kit. Tesla dreamed up the alternating-current motor during a stroll through the Városliget, or city park, in Budapest.[44] To ponder difficult problems, Darwin walked down a gravel path called the "sandwalk" near his home in Kent, kicking up stones along the way.[45] The physicist Werner Heisenberg devised the uncertainty principle during a late-night walk through a park in Copenhagen.[46] For two years, he had been frustrated that his equations could predict the momentum of a quantum particle but not its position. One night, he had an epiphany: What if there was no problem with the equations? What if the uncertainty was actually inherent in the nature of quantum particles? After walking with the question long enough, Heisenberg gradually walked into the answer.

Some scientists turn to music to tap into their subconscious. Einstein, for example, played his violin to decipher the music of the cosmos. As one friend recalled, "He would often play his violin in his kitchen late at night, improvising melodies while he pondered complicated problems. Then suddenly, in the middle of playing, he would announce excitedly, 'I've got it!' As if by inspiration, the answer to the problem would have come to him in the midst of music."[47]

Many creative people also embrace idleness to spur original thought. Ideas "come from daydreaming," the author Neil Gaiman explains. They come "from drifting, that moment when you're just sitting there." When people ask Gaiman for advice on how to be a writer, his answer is simple: "Get bored."[48] Stephen King agrees: "Boredom can be a very good thing for someone in a creative jam."[49]

Getting bored landed a woman named Joanne her first publishing deal. In 1990, her train from Manchester to London was delayed for four hours. While waiting for the train, a story "came fully formed" into her mind—about a young boy who attends a wizardry school.[50] That four-hour delay ended up being a blessing for Joanne "J. K." Rowling, whose Harry Potter series captivated millions around the world.

In one sense, Rowling was lucky. Her epiphany arrived before smartphones did, so she didn't have to play defense against notifications while waiting for her train. But the rest of us have to be proactive about building boredom into our lives. Bill Gates, for example, goes to a secluded Pacific Northwest cabin for weeklong retreats that he calls "Think Week" dedicated—you guessed it—to thinking without distractions.[51] Phil Knight, the cofounder of Nike, had a designated chair in his living room for daydreaming.[52]

Following in their footsteps, I decided to break my codependency with my phone and proactively rekindle my long-lost affair with boredom. I began deliberately building time into my day—an airplane mode of sorts—when I sit on my recliner doing nothing but thinking. I spend twenty minutes, four days a week, in the sauna, with nothing but a pen and paper in hand. Odd place for writing? Yes. But some of the best ideas in recent memory occurred to me in that solitary, stifling environment.

It sounds so simple. A walk in the park. A shower. Sitting in the sauna or a chair to daydream. But there's no magic here—at least not in the Hogwarts sense. The magic is the intention of a designated time to pause and reflect—a moment for interior silence to oppose contemporary chaos.

In an age of instant gratification, this habit can sound a bit underwhelming. But creativity often comes as a subtle whisper—not a big bang. You must be patient enough to pursue the whisper and perceptive enough to receive it when it arrives. If you live with the question long enough, "you will gradually, without noticing it, one distant day live right into the answer," as the poet Rainer Maria Rilke wrote.[53]

The next time you feel boredom arising, resist the temptation to take a hit of data or do something "productive." Boredom might just be the most productive thing you can do.

Boredom has another benefit. It allows your mind to freely associate and draw connections between drastically different objects—say, an apple and an orange.

Comparing Apples and Oranges

Many idioms in English have flummoxed me since I started learning the language in middle school. But one tops the list: comparing apples and oranges. The first time I heard the idiom in college, it stopped me in my tracks. I thought there's more that unites apples and oranges than divides them. (At this point, dear reader, you may want to turn around and look away. I'm about to compare apples and oranges.) Both are fruits. Both are round(ish). Both have a slight tangy taste. Both are about the same size. And both grow on trees.

Scott Sanford of the NASA Ames Research Center took this comparison a step further. He used infrared spectrometry to compare a Granny Smith apple and a navel orange and showed that the spectra of the two fruits are strikingly similar. The study, with the tongue-in-cheek title "Apples and Oranges: A Comparison," was published in the satirical scientific magazine *Improbable Research*.[54]

Despite the similarities between apples and oranges, the idiom thrives because we're terrible at seeing connections between seemingly dissimilar or unrelated things. In our personal and professional lives, we confine ourselves to comparing apples to apples and oranges to oranges.

Specialization is all the rage these days. In the English-speaking world, a generalist is a Jack or Jill of all trades, but the master of none. The Greeks caution that a person "who knows a lot of crafts lives in an empty house."[55] The Koreans believe a person of "12 talents has nothing to eat for dinner."[56]

This attitude comes at a cost. It stifles the cross-pollination of ideas from different disciplines. We remain in our humanities track or science track and shut off our minds to concepts from across the aisle. If you're an English major, what use do you have for quantum theory? If you're an engineer, why bother reading Homer's *Odyssey*? If you're a medical student, why study the visual arts?

That last question was the subject of a research study.[57] Thirty-six first-year medical students were randomly split into two groups. The first group took six classes at the Philadelphia Museum of Art on observing, describing, and interpreting works of art. These students were compared with a control group that didn't enroll in the art classes. Unlike the control group, the art-training group members significantly improved their observational skills—such as interpreting photographs of retinal disease—as measured in tests conducted at the beginning and the end of the study. "Art training *alone*," the study suggests, "can help to teach medical students to become better clinical observers."[58]

Life, it turns out, doesn't happen in compartmentalized silos. There's little to be learned from comparing similar things. "To create," biologist François Jacob said, "is to recombine."[59] Decades later, Steve Jobs echoed the same sentiment: "Creativity is just connecting things. When you ask creative people how they did something, they feel a little guilty because they didn't really *do* it, they just *saw* something. . . . [T]hey've had more experiences or they have thought more about their experiences than other people."[60]

Put differently, it's easier to "think outside the box" when you're playing with multiple boxes.

Einstein called this idea "combinatory play," which he believed is the "essential feature in productive thought."[61] Combinatory play requires exposing yourself to a motley coalition of ideas, seeing the similar in the dissimilar, and combining and recombining apples and oranges into a brand-new fruit. With this approach, the "whole becomes not only more than but very different from the sum of its parts," as the physicist and Nobel laureate Philip Anderson explains.[62]

To facilitate cross-pollination, renowned scientists often develop diverse interests. Galileo, for example, was able to spot mountains and plains on the Moon—not because he had a superior telescope, but because his training in painting and drawing enabled him to understand what the bright and dark regions on the Moon represented.[63] Leonardo da Vinci's inspiration for art and technology also came from the outside—in his case, nature. He instructed himself on subjects as diverse as "the placenta of a calf, the jaw of a crocodile, the tongue of a woodpecker, the muscles of a face, the light of the moon, and the edges of shadows."[64] Einstein's inspiration for general relativity came from the eighteenth-century Scottish philosopher David Hume, who first questioned the absolute nature of space and time. In a December 1915 letter, Einstein wrote, "It is very possible that without these philosophical studies I can not say that [relativity] would have come."[65] Einstein was first introduced to Hume's work through a group called the Olympia Academy, a group of friends dedicated to combinatory play who met in Bern, Switzerland, to discuss physics and philosophy.

In developing his theory of evolution, Darwin was inspired by two very different fields—geology and economics. In *Principles of Geology*, Charles Lyell argued in the 1830s that mountains, rivers, and canyons had been formed through a slow, evolutionary process that took place over eons as erosion, wind, and rain chipped away at the Earth. Lyell's theory bucked conventional wisdom, which attributed these geological features solely to catastrophic or supernatural events like Noah's flood.[66] Darwin read Lyell's book while sailing on the *Beagle* and applied the geological idea to biology. As rocket scientist David Murray explains, Darwin argued that organic material "evolves just as inorganic material does: with minute changes in each descendant that, over time, accumulate to form new biological appendages like eyes, hands, or wings."[67] Darwin also drew inspiration from the late-eighteenth-century economist Thomas Malthus. Malthus argued that humans tend to outgrow resources like food, creating a competition for survival. This competition, Darwin believed, drove the evolutionary process, leading the species best adapted to their environment to survive.[68]

Combinatory play is also the hallmark of great musicians. The renowned record producer Rick Rubin tells his bands not to listen to popular songs while they produce an album. They're "better off drawing inspiration from the world's greatest museums," Rubin says, "than finding it in the current Billboard charts."[69] For example, Iron Maiden's music combines the unlikely elements of Shakespeare, history, and heavy metal. Queen's "Bohemian Rhapsody," considered one of the greatest rock songs of all time, is like a musical sandwich, blending an opening and closing ballad with hard rock and opera in the middle.

David Bowie was another master blender. When writing lyrics, he used a custom-developed computer program called the Verbasizer.[70] Bowie would type up sentences from different sources—newspaper articles, journal entries, and the like—into the Verbasizer, which would cut them up into words and mix and match them. "What you end up with," Bowie explained, "is a real kaleidoscope of meanings and topic[s] and nouns and verbs all sort of slamming into each other." These combinations would then serve as inspiration for song lyrics.

Combinatory play has also produced many breakthrough technologies. Larry Page and Sergey Brin adopted an idea from academia—the frequency of citations to an academic paper indicates its popularity—and applied it to the search engine to create Google. Steve Jobs famously borrowed from calligraphy to create multiple typefaces and proportionally spaced fonts on the Macintosh. Netflix cofounder Reed Hastings was inspired by the subscription model used at his gym: "You could pay $30 or $40 a month and work out as little or as much as you wanted."[71] Frustrated by the big late fees he had incurred for renting *Apollo 13*, Hastings decided to apply the same model to video rentals.

Nike's first running shoes were modeled after a common household appliance.[72] In the early 1970s, University of Oregon running coach Bill Bowerman was looking for shoes that would perform well on different surfaces. At the time, Bowerman's athletes would wear shoes with metal spikes, which lacked proper traction and would destroy the running surface.

One Sunday morning over breakfast, Bowerman's eyes drifted toward an old waffle iron in the kitchen. He spotted the gridded pattern on the waffle iron and thought that by turning the pattern upside down, he could create a shoe without spikes. He grabbed the waffle iron, took it to his garage, and began to create molds. What came out of these experiments was the Nike Waffle Trainer, a revolutionary shoe with rubber traction that provided better grip and adapted to the running surface. The original waffle iron from Bowerman's kitchen now sits in a display case at Nike's headquarters.

As these examples show, a revolution in one industry can begin with an idea from another. In most cases, the fit won't be perfect. But the mere act of comparing and combining will spark new lines of thinking.

We can't combine ideas if we don't see the similarities between them. Biologist Thomas H. Huxley, after reading *On the Origins of Species*, purportedly said, "How extremely stupid [of me] not to have thought of that!"[73] The connection between apples and oranges appears obvious—but only in hindsight. In Darwin's time, there were many people who had studied species. There were also many people who read Malthus and Lyell, the economist and the geologist who had inspired Darwin. But it was the rare person who studied species, who read Malthus, *and* who read Lyell—and who could make the connections between the three fields.

As these examples show, to connect apples and oranges, you have to collect them first. The more diverse your collection, the more interesting your output becomes. Pick up a magazine or book about a subject you know nothing about. Attend a different industry's conference. Surround yourself with people from different professions, backgrounds, and interests. Instead of talking about the weather and repeating other small-talk platitudes, ask, "What's the most interesting thing you're working on right now?" The next time you find yourself in a creative jam, ask, "What other industry has faced an issue like this before?" For example, Johannes Gutenberg had a printing press problem, so he looked to other industries—like winemakers and olive oil producers—who used a screw press to extract juice and oil. Gutenberg then applied the same concept to kick-start the era of mass communication in Europe.

Organizations can take a cue from Pixar, the creative studio behind numerous box-office hits, such as *Toy Story* and *Finding Nemo*. The company encourages its employees to spend up to four hours a week taking classes at Pixar University, its professional-development program. The classes include painting, sculpting, juggling, improv, and belly dancing.[74] Although these classes don't have a direct bearing on filmmaking, Pixar knows that creative ideas come from seemingly unlikely places. If you keep collecting apples and oranges and spending time with them, ideas for new fruits will begin arriving soon enough.

The principle of combinatory play applies not just to ideas, but to people as well. As we'll see in the next section, when people from different disciplines are combined, the result is more than the sum of their parts.

The Myth of the Lone Genius

"These rovers are so complicated that no one understands them."

This might strike you as an odd statement coming from Steve Squyres, the principal investigator of the 2003 Mars Exploration Rovers project. He led the team responsible for dreaming up the rovers, devising the on-board instruments, and operating them on the Martian surface. But even to Squyres, the rovers are "too complicated for a single person to wrap their head around completely." The understanding comes not individually, but as part of the collective brain.

We often fetishize the lone genius toiling away in the garage—whether it's Bowerman tinkering with his waffle iron in his own garage or Jobs building the first Apple computer from the garage of his family home. It makes for a great story, but, like most stories, it's a misleading depiction of how things work.

Optimal creativity doesn't happen in complete isolation. Breakthroughs almost always involve a collaborative component. "If I have seen further," Newton famously said, "it is by standing upon the shoulders of giants." These giants come to the table with diverse perspectives, bringing their own apples and oranges for the collective body to compare and connect.

Entrepreneur and writer Frans Johansson calls this phenomenon the Medici effect. It refers to the fifteenth-century creative explosion that occurred when the wealthy Medici family brought together in Florence many accomplished individuals from different walks of life—scientists, poets, sculptors, philosophers, and more. As these individuals connected, new ideas blossomed, paving the way for the Renaissance (the word means "rebirth" in French).[75]

A mission to Mars produces its own Medici effect by bringing together scientists and engineers to collaborate on the mission. Although these two groups tend to be lumped together in popular accounts of space exploration, they belong to very different tribes.[76] Scientists are the idealistic truth seekers trying to understand how the universe works. The engineers, on the other hand, are more pragmatic. They must design hardware capable of implementing the scientists' vision, while grappling with practical realities like finite budgets and schedules.

Opposites don't always attract. On each mission, there's a tension between "the idealistic, impractical scientists" and "the stubborn, practical engineers," as Squyres writes. On the good missions, this tension turns into a creative dance that brings out the best in both disciplines. But on the bad missions, "it's an acid that eats away at the collaboration until it's rotten."[77]

The key to making the relationship work is combinatory play. The scientists learn some engineering, and the engineers learn some science. This approach was a top priority for Squyres. "If you came in," he explains, "and you sat in on one of our daily tactical planning sessions where we have a team of a dozen scientists and a dozen engineers sitting together in a room, you could sit there for an hour and still not quite figure out who were the scientists and who were the engineers." The team was blended so well—with the scientists and the engineers well versed in each other's language and objectives—that you could hardly tell the difference.

You might assume that today's work environment is an ideal setting for this type of blending. Sitting in cubicles in open offices, and connected through always-on technologies like email and Slack, most modern workers are constantly collaborating with each other. Maybe it's time for a new renaissance of a modern sort that will be dubbed the Slack effect.

But not so fast. Consider the result of one study, where researchers separated the participants into three groups and asked them to solve a complex problem.[78] The first group worked in complete isolation, the second group was in constant interaction, and the third group alternated between interaction and isolation.

The best-performing group was the third. "Intermittent breaks in interaction improve collective intelligence," the researchers observed.[79] Cycling between isolation and interaction improved the average score of the group while also leading the group to find the best solutions more frequently. Importantly, both low performers and high performers in the group benefited from intermittent interaction. These results suggest that learning flowed in both directions, with one person's conclusions becoming input for the others.[80]

Most modern work environments resemble the second, constant-interaction group, an arrangement suboptimal for creativity. As the research shows, connection is important, but so is time for isolated reflection. The process of creation can be embarrassing. "For every new good idea you have," Asimov writes, "there are a hundred, ten thousand foolish ones, which you naturally do not care to display."[81] People should be able to cultivate insights on their own, come together to exchange those insights with the group, and then return to working alone, cycling between solitude and collaboration. The pattern is similar to the focus and boredom cycle that we explored earlier.

When it comes to boosting creativity, cognitive diversity—blending together your version of scientists and engineers—isn't just a buzzword. It's a necessity. But there's another level of cognitive diversity that often gets overlooked.

Beginner's Mind

In the 1860s, the silk industry in France was endangered by a disease that threatened silkworms. Chemist Jean-Baptiste Dumas urged his former student, Louis Pasteur, to work on the problem. Pasteur was hesitant. "But I never worked with silkworms," he protested. Dumas replied, "So much the better."[82]

Most of us don't do what Dumas did. We instinctively dismiss the opinions of amateurs like Pasteur. *They don't know what they're talking about. They haven't attended the relevant meetings. They don't have the necessary background. They're out of their element.*

Yet it's precisely for these reasons that outsider opinions hold value.

First-principles thinking, as Dumas's answer implies, often has an inverse relationship to expertise. Unlike the insiders, whose identity or salary can depend on the existing state of affairs, outsiders have no stake in the status quo. Conventional wisdom is easier to tune out when you're not smothered by it.

Consider, for example, the geological theory of continental drift, which says that continents were one big mass and drifted apart over time. The theory is the brainchild of Alfred Wegener—a meteorologist and an outsider to geology.[83] Continental drift was initially declared absurd by geological experts who assumed that continents were stable and didn't move. Geologist R. Thomas Chamberlain summed up the collective sentiments of the insiders: "If we are to believe in Wegener's hypothesis we must forget everything which has been learned in the past 70 years and start all over again."[84] Wegener's theory would upend the foundations of the insiders' reputation in the field, so they stuck to their guns. For similar reasons, when Johannes Kepler discovered that planets had elliptical—rather than circular—orbits, Galileo balked. As astrophysicist Mario Livio observes, "Galileo was still prisoner to the aesthetic ideals of antiquity, which assumed that the orbits had to be perfectly symmetrical."[85]

Einstein's secret to success was escaping the intellectual prison that confined other physicists. When he published his paper on special relativity, he was an unknown clerk in a Swiss patent office. As an outsider to the physics establishment, he was able to move beyond the collective body of knowledge—which, in his case, was a Newtonian perspective that treated time and space as absolute. His revolutionary paper on special relativity, "On the Electrodynamics of Moving Bodies," looks nothing like a typical physics paper. It cites the names of only a handful of scientists and contains virtually no citations to existing works—a highly unconventional move by academic standards.[86] In Einstein's case, creating a revolution meant searching beyond incremental improvement, untethered by citations to works past.

Other examples abound. Musk was a latecomer to rocket science, which he picked up by reading textbooks. Bezos came to retail from the finance world, and Hastings was a software developer before he cofounded Netflix. Standing outside the establishment, these gate-crashers were in a better position to see its flaws and recognize its outdated methods.

In Zen Buddhism, this principle is known as *shoshin*, or beginner's mind.[87] As the Zen teacher Shunryu Suzuki writes, "In the beginner's mind there are many possibilities; in the expert's mind there are few."[88] This is why Wieden+Kennedy, the advertising firm responsible for many of Nike's blockbuster ad campaigns, encourages its employees to "walk in stupid" every day and approach problems from a beginner's viewpoint.

It was a beginner who created a billion-dollar author. When J. K. Rowling submitted a draft copy of the first Harry Potter book to publishing houses, they were unanimous in their opinion: They thought the book was not worth printing. Her manuscript was rejected by numerous publishers, until it ended up on the desk of Nigel Newton, the chairman of Bloomsbury Publishing.[89] Newton saw potential in the book where his rivals missed it.

How? His secret was his eight-year-old bookworm daughter, Alice.[90] After Newton handed a sample from the book to Alice, she devoured it and nagged him for more. "Dad," she said, "this is so much better than anything else." Alice's input convinced her father to write a £2,500 check to Rowling as a meager advance for acquiring the rights to publish her book. The rest is history.

What gave Newton the multimillion-pound edge was his willingness to get the opinion of his daughter—an outsider to the publishing industry, but a member of the target audience for the book.

This isn't to suggest that all original ideas come from beginners. To the contrary, expertise is valuable in idea generation, but experts shouldn't work in complete isolation, the lone genius lore be damned. Experts also benefit from intermittent periods of collaboration, particularly when amateurs are brought into the mix.

....

IT DOESN'T TAKE a genius polymath to devise thought experiments. All it takes is a desire to collect apples and oranges, the patience to sit in boredom while your subconscious compares and connects them, and a willingness to expose the new fruits to others—whether it's the scientists on your engineering team or your eight-year-old daughter.

Now that we've gotten more comfortable with thought experiments, it's time to turn up the volume on your imagination and start reaching for the Moon.

Visit **ozanvarol.com/rocket** to find worksheets, challenges, and exercises to help you implement the strategies discussed in this chapter.

4

MOONSHOT THINKING

The Science and Business of the Impossible

ALICE: There's no use trying. One can't believe impossible things.
WHITE QUEEN: I daresay you haven't had much practice. When I was
your age, I always did it for half-an-hour a day. Why, sometimes I've
believed as many as six impossible things before breakfast.
—LEWIS CARROLL, *Through the Looking-Glass*

CHARLES NIMMO WAS an unlikely choice for a test subject.[1] A sheep farmer in the small, rural town of Leeston, New Zealand, he volunteered to take part in a secret project involving the flight of a secret object. During earlier test flights in California and Kentucky, the object had been mistaken by numerous observers for a UFO. It was picked up by CNN and generated headlines in local newspapers—"Mystery Object in Sky Captivates Locals," reported the *Appalachian News-Express*.[2]

Nimmo is one of over four billion people in the world who lack access to a technology many of us take for granted: high-speed internet. The internet is as revolutionary as the electrical grid. Once you're plugged in, you can power up your life. According to a Deloitte study, bringing reliable internet access to Africa, Latin America, and Asia would "generate more than $2 trillion in additional GDP."[3] Internet access can lift people out of poverty, save lives, and, in Nimmo's case, provide access to information about the weather, which is crucial for a sheep farmer. Nimmo needs to know when his sheep will be sufficiently dry for crutching—a technical term for shearing the wool from a sheep's butt.

Lighting up the world with cheap, reliable internet access isn't easy. Satellite-powered internet is expensive and produces weak signals with a significant transmission delay, given the distance a signal must travel to and from a satellite in Earth orbit. Land-based cell towers often have limited ranges and don't make economic sense for many rural, sparsely populated areas—even in developed countries like New Zealand. Challenging geography, like mountains and jungles, can also prevent cell tower signals from reaching their destinations.

Nimmo was the first test subject for an audacious project intended to lift the internet blackout that covers much of the world. The project is the brainchild of X, formerly known as Google X. The notoriously secretive company is dedicated to researching and developing breakthrough technologies. X doesn't innovate *for* Google. X creates the next Google.

To solve the internet access problem, Xers (as they're called) came up with a loony thought experiment: What if we used balloons?

They imagined balloons the size of tennis courts, shaped like giant jellyfish, hovering in the stratosphere at around sixty thousand feet—above the weather and air traffic. The balloons would carry small computers in polystyrene boxes, powered by solar energy to beam internet signals down to the ground.

You might be wondering why a story about ballooning—a rather primitive technology—appears in this book. Ballooning, after all, isn't rocket science. In fact, ballooning "is way harder than rocket science," says a former Xer. Because balloons get pushed around easily by the winds, they must be steered like a sailboat to catch the right air currents. Reliable connectivity is also difficult to achieve when the balloons are constantly moving around.

X's solution to this problem was to create a network of balloons that would work together like a daisy chain and ensure reliable connectivity.

When one balloon left, another would take its place. The balloons would live several months before landing back on Earth to be reused.

This loony project got an appropriately loony title: Project Loon. After delivering internet access to Nimmo the sheep farmer and conducting other test missions, the Loon balloons went on to fly more than thirty million miles. When catastrophic floods hit Peru in early 2017, the balloons came to the rescue. The floods affected hundreds of thousands of people and knocked out the communications network across the country. In less than seventy-two hours, Project Loon showed up on the scene and began delivering basic connectivity to tens of thousands of Peruvians.[4] Later that year, after Hurricane Maria devastated Puerto Rico, Loon distributed help in the form of balloon-powered internet to the hardest-hit parts of the island.[5]

Loon was a moonshot—a breakthrough technology that brings a radical solution to an enormous problem. This chapter is about the power of moonshot thinking, which is responsible for audacious projects like Loon. We'll explore why some of the greatest achievements in history have their roots in moonshot thinking. I'll explain why you should act more like a fly and less like a bee, and why you're better off hunting for antelopes instead of mice. You'll discover how the use of a single word can boost creativity, what you should do first in tackling an audacious goal, and why sketching a path to the future often requires moving back from it.

The Power of Moonshot Thinking

The Moon is our most ancient companion. It has kept us company for much of the Earth's existence. As Robert Kurson writes, the Moon has "controlled tides, guided the lost, lit harvests, inspired poets and lovers, spoke[n] to children."[6] And since our ancestors first gazed their heads upward, the Moon has tantalized us, appealing to a primal instinct to explore beyond our home. But for most of our existence, it has remained a moonshot, far out of our reach.

When President Kennedy gave the speech that opened this book—where he looked to the future and picked the Moon as our new frontier—it appeared that he was hoping for a miracle. Kennedy asked his nation "to do what most people thought was impossible," as Apollo astronaut Gene Cernan recalled, "including me."[7] The promise to put a human being on the Moon in less than a decade was so incredible, remembers Robert Curl, a Rice University professor who was in the audience for Kennedy's speech, "I came away in wonder that he was seriously proposing this."[8]

Famous NASA flight director Gene Kranz—who was played by Ed Harris in the movie *Apollo 13*—was also stunned by Kennedy's bold pledge.[9] For Kranz and his NASA colleagues "who had watched [their] rockets keel over, spin out of control, or blow up, the idea of putting a man on the Moon seemed almost too breathtakingly ambitious."[10] But Kennedy was well aware of the difficulties ahead. "We choose to go to the moon in this decade and do the other things," Kennedy said, "not because they are easy, but because they are hard." He simply refused to let the existing reality drive his country's future.

This was humanity's first actual moonshot. But humans had been taking metaphorical moonshots long before Neil Armstrong and Buzz Aldrin walked on the Moon. When our ancestors blazed a trail to some unknown corner of the earth, they took a moonshot. The discoverers of fire, the inventors of the wheel, the builders of the pyramids, the makers of automobiles—they all took moonshots. It was a moonshot for slaves to reach for freedom, for women to take the ballot, and for refugees to push toward distant shores in search of a better life.

We're a species of moonshots—though we've largely forgotten it.

Moonshots force you to reason from first principles. If your goal is 1 percent improvement, you can work within the status quo. But if your goal is to improve tenfold, the status quo has to go. Pursuing a moonshot puts you in a different league—and often an entirely different game—from that of your competitors, making the established plays and routines largely irrelevant.

Here's an example.[11] If your goal is to improve car safety, you can make gradual improvements to the design of a car to better protect human life in an accident. But if your goal is a moonshot of eliminating all accidents, you must start with a blank slate and question all assumptions—including the human operator behind the wheel. This first-principles approach paves the way for the possibility of autonomous vehicles.

Consider also the planned moonshots of SpaceX. If the company's aim were to simply put satellites into Earth orbit, there would have been no reason to do things differently. The company would have relied on the same technology that NASA had been using since the 1960s. There's little reason to reduce the cost of rocket launches by a factor of ten, as SpaceX is on its way to doing, unless you're aiming for a moonshot. The bold ambition of colonizing Mars forced SpaceX to employ first-principles thinking and transform the status quo.

The political strategists James Carville and Paul Begala tell a story about the choice a lion faces in deciding to hunt for a mouse or an antelope. "A lion is fully capable of capturing, killing, and eating a field mouse," they explain. "But it turns out that the energy required to do so exceeds the caloric content of the mouse itself." Antelopes, in contrast, are much bigger animals, so "they take more speed and strength to capture." But once captured, an antelope can provide days of food for the lion.[12]

The story, as you may have guessed, is a microcosm for life. Most of us go after the mice instead of the antelopes. We think the mouse is a sure thing, but the antelope is a moonshot. Mice are everywhere; antelopes are few and far between. What's more, everyone around us is busy hunting mice. We assume that if we decide to go for antelopes, we might fail and go hungry.

So we don't launch a new business, because we think we don't have what it takes. We hesitate to apply for a promotion, assuming that someone far more competent will get it. We don't ask people on a date if they seem out of our league. We play not to lose instead of playing to win. "The story of the human race," psychologist Abraham Maslow wrote in 1933, "is the story of men and women selling themselves short."[13]

If Kennedy were following this mindset, his speech would have been very different (and far more boring). "We choose," he might have said, "to put humans in Earth orbit and make them circle round and round—not because it's challenging—but because it's doable given what we have." (Which, incidentally, is exactly what NASA decided to do in the 1980s. More on that later.)

Setting your sights low is the moral of the Icarus myth. Icarus's father, the craftsman Daedalus, built wings out of wax for himself and his son to escape the island of Crete. Daedalus warned his son to follow his flight path and not to fly too close to the Sun. You probably know what happened next: Icarus ignored his father's warnings and soared near the Sun. His wings melted, sending Icarus on a fatal plunge into the sea.

The lessons of the myth are clear: Those who soar melt their wings and die. Those who follow the predefined path and obey instructions escape the island and survive.

But as Seth Godin explains in his book *The Icarus Deception*, there's a second half to the Icarus myth—one that you probably haven't heard. In addition to telling Icarus not to fly too high, Daedalus also told him not to fly too *low*, because the water would ruin his wings.[14]

Altitude, as any pilot will tell you, is your friend. If your engine quits when you're flying high, you've got options for gliding your plane to safety. But at low altitudes, the possibilities in flight—like the possibilities in life—are more limited.

Businesses that fly at higher altitudes tend to perform better. Shane Snow summarizes the relevant research in *Smartcuts*: "From 2001 to 2011, an investment in the 50 most idealistic brands—the ones opting for the high-hanging purpose and not just low-hanging profits—would have been 400 percent more profitable than shares of an S&P index fund."[15] Why? Moonshots appeal to human nature and attract more investors. Poking fun at the limited ambitions of most Silicon Valley firms, the manifesto for Founders Fund—a prominent venture-capital firm—reads: "We wanted flying cars, instead we got 140 characters."[16] The firm became the first outside investor in SpaceX's moonshots.

Moonshots are also talent magnets. This is why SpaceX and Blue Origin have been able to cherry-pick the best rocket scientists from traditional aerospace companies and make them work around the clock on audacious engineering projects. Musk's selling point was that the engineers would "have the freedom to actually do their job—build a rocket—rather than sitting in daylong meetings, waiting months for a parts request to wend its way through a bureaucracy, or fending off internal political attacks."[17]

You might be thinking, it's easy for internet billionaires to start a space company. It's easy for Kennedy to pursue a moonshot when Congress has funneled billions of dollars into beating the Soviets to the Moon. It's easy for X, backed with the financial might of Google, to pursue outlandish ideas like Project Loon. But, you might think, it's impossible to chase moonshots when you've got a business to keep afloat, mortgage payments to make, and board members to please.

This is an objection that Astro Teller—the captain of moonshots at X (yes, that's his real title)—hears frequently. "Somehow society has developed this notion that you have to have a huge amount of money to be audacious," he says. But Teller doesn't buy it: "Taking good, smart risks is something that anyone can do, whether you're on a team of 5 or in a company of 50,000."[18] Bezos would agree. "Given a ten percent chance of a 100 times payoff, you should take that bet every time," he wrote in Amazon's annual letter to shareholders in 2015. But most of us won't place bets that have even a 50 percent chance of success, regardless of the potential reward.

Yes, some moonshots *are* too impractical to materialize in the near future—if ever. But you don't need all your moonshots to take flight. As long as your portfolio of ideas is balanced—and you're not betting your future on a single moonshot—one big success will compensate for the ideas better left to novels and movies. "If you place enough of those bets," Bezos says, "and if you place them early enough, none of them are ever betting the company."[19]

Here's the thing: The hurdle to taking moonshots isn't a financial or practical one. It's a mental one. "Not many people believe that they can move mountains," David Schwartz says in *The Magic of Thinking Big*. "So, as a result, not many people do."[20] The primary obstacles to moonshots are in your head, reinforced by decades of conditioning by society. We've been seduced into believing that flying lower is safer than flying higher, that coasting is better than soaring, and that small dreams are wiser than moonshots.

Our expectations morph reality and become self-fulfilling prophecies. What you strive for becomes your ceiling. Go for mediocrity, and mediocrity is what you'll get—at best. You can't always get what you want, as the Rolling Stones remind us. But if you course-correct in the direction of the Moon—as opposed to the ground—you'll soar higher than you would have before. "If you set your goals ridiculously high and it's a failure, you will fail above everyone else's success," says James Cameron, the filmmaker behind such blockbusters as *The Terminator* and *Titanic*.[21]

Many of us refrain from moonshots because we assume we're not cut out for them. We believe the kind of people who can fly high have better wings impervious to melting. Michelle Obama dispelled this myth in a 2018 interview. "I have been at probably every powerful table that you can think of," she explained, "I have worked at nonprofits, I have been at foundations, I have worked in corporations, served on corporate boards, I have been at G-summits, I have sat in at the UN: They are not that smart."[22]

They are not that smart. They just know what most of us have never learned: There's far less competition for antelopes. Everyone else is busy chasing mice in the same crowded, rapidly shrinking territory. This means you can't afford *not* to take moonshots. If you wait too long—if you keep chasing ever-smaller business margins at ever-greater cost—someone else will take the moonshot that puts you out of a job or makes your business obsolete.

The story we choose to tell ourselves about our capabilities is just that: a choice. And like every other choice, we can change it. Until we push beyond our cognitive limits and stretch the boundaries of what we consider practical, we can't discover the invisible rules that are holding us back. There are tremendous benefits to taking moonshots even where—or particularly where—real-life conditions are out of sync with our imagination.

Take comfort in knowing that Daedalus had his physics all wrong. Air gets cooler, not hotter, as you ascend, so your wings won't melt. If you pursue the extraordinary, you'll rise above the stale neural pathways that dominate ordinary thinking. And if you persist—and learn from the inevitable failures that will arise—you'll eventually grow the wings you need to soar.

Growing those wings requires a strategy called divergent thinking, which we'll explore in the next section.

Embracing the Far-Fetched

Imagine a glass bottle with its base pointed toward a light. If you put half a dozen bees and flies into the bottle, which species would find its way out first?

Most people assume the answer is bees. After all, bees are known for their intelligence. They can learn highly complex tasks—such as lifting or sliding a cap to access a sugar solution in a lab—and teach what they learned to other bees.[23]

But when it comes to finding their way out of the bottle, the bees' intelligence gets in their way. The bees love the light. They'll keep bumping up against the base of the bottle—located at the light source—until they die of exhaustion or hunger. In contrast, the flies disregard "the call of the light," as Maurice Maeterlinck writes in *The Life of the Bee*. They "flutter wildly hither and thither" until they stumble on the opening at the other end of the bottle that restores their liberty.[24]

The flies and the bees, respectively, represent what's known as divergent and convergent thinking. The flies are the divergent thinkers, fluttering freely until they find the exit. The bees are the convergent thinkers, zeroing in on the seemingly most obvious exit path with a behavior that is ultimately their undoing.

Divergent thinking is a way of generating different ideas in an open-minded and free-flowing manner—like flies bouncing around in a glass bottle. During divergent thinking, we don't think about constraints, possibilities, or budgets. We just throw around ideas, open to whatever might present itself. We become optimists in the way that physicist David

Deutsch defines the term—as someone who believes that anything permitted by the laws of physics is doable.[25] The goal is to create a flurry of options—both good and bad—not prematurely judging them, limiting them, or choosing among them.

At the initial stages of idea formation, "the pure rationalist has no place," as the physicist Max Planck put it. Discovery, as Einstein also explained, "is not a work for logical thought, even if the final product is bound in logical form."[26] To activate divergent thinking, you must shut down the rational thinker in you, the part responsible for otherwise safe, beneficial grown-up behaviors. Set aside the spreadsheets, and let your brain run wild. Investigate the absurd. Reach beyond your grasp. Blur the line between fantasy and reality.

Research shows that divergent thinking is a portal to creativity. It boosts people's ability to discover innovative solutions and make new associations. In other words, it lets you compare and connect apples and oranges.[27]

Consider a study by three Harvard Business School professors who gave the participants a difficult ethical challenge.[28] The researchers laid out a scenario where the ethical choice wasn't obvious and divided the study's participants into groups. To one group, they asked, "What *should* you do?" To the other group, they asked, "What *could* you do?" The "should" group zeroed in on the most obvious solutions—often not the best ones—but the "could" group stayed open-minded and generated a broader range of possible approaches. As the researchers explained, "People may often benefit from a *could* mindset that involves a more expansive exploration of possible solutions before making a final decision." A different study reached the same conclusion. Participants who were told "Object A *could* be a dog's chew toy" as opposed to "Object A *is* a dog's chew toy" generated a greater variety of uses for the toy.[29]

It's tempting to skip divergent thinking and instead resort to convergent thinking—evaluate what's easy, what's probable, what's doable. Convergent thinking is like taking a multiple-choice exam: You pick from a limited, predetermined menu of options with no ability to write in a new answer. You assume, as the bees did, that there's only one way out—flying toward the light. As Stanford business professor Justin Berg writes, "Convergent thinking alone is dangerous because you're just relying on the past. What will succeed in the future may not resemble what succeeded in the past."[30]

To test this idea, Berg conducted a study of Cirque du Soleil performers.[31] He evaluated the roles played by creators, who produce ideas for new circus acts, and managers, who decide whether to include them in the shows. He found that managers performed abysmally in predicting the success of new circus acts. They relied too heavily on convergent thinking, preferring conventional performances over novel acts. Although creators overestimated the promise of their own ideas, they were far more accurate than managers in judging the creative promise of their colleagues' novel acts. Their ability to think divergently—coupled with their distance from the ideas—gave them a significant edge.

Divergent thinking does *not* mean thinking happy thoughts, sprinkling some pixie dust, and watching them take flight. We need the idealism of divergent thinking to be followed by the pragmatism of convergent thinking. "The creative process is not about one state," science historian Steve Johnson explains. "It's the ability to move between different mental states."[32] Recall from earlier that cycling between moments of solitude and moments of collaboration creates the optimal environment for creativity. It's a similar idea here. You should cycle between a fly mindset and a bee mindset, but you've got to do things in the right order. We have to generate ideas first before we can begin evaluating and eliminating them. If we cut the accumulation process short—if we immediately start thinking about consequences—we run the risk of hampering originality.

We've all been in that meeting before. People are gathered around a conference table, with half-empty cups of lukewarm coffee strewn around, to "brainstorm ideas" and "explore options." But instead of exploring ideas, everyone's busy shooting them down. "We've tried that before." "We don't have the budget." "The management would never approve." Idea generation stops before it even begins. As a result, instead of trying something new, we end up doing what we did yesterday. The goal should be to re-sist the tendency to activate convergent thinking through a "This can't be done" attitude. Instead, begin with a divergent "This *could* be done if . . ." mindset.

We know surprisingly little about how the brain works, but according to one theory, idea generation and idea evaluation take place in different parts of the brain.[33] For example, researchers in a University of Haifa study used a functional magnetic resonance imaging (fMRI) machine to evalu-ate how much oxygen different parts of the brain consume during creative tasks. They found that individuals who were more creative had decreased activity in the sections of the brain associated with evaluation.[34]

Because of the differences between idea generation and idea evalua-tion, many authors separate their drafting from their editing. Drafting is better suited for divergent thinking, and editing for convergent. During my research for this book, I collected vast amounts of information from any source I could find. I adopted a broad definition of *relevant*, erring on the side of overinclusion and bouncing from one part of the bottle to the other. I applied a similar approach in writing the book's first draft—not overthinking matters like structure, etiquette, or even proper grammar—just putting down one crappy sentence after the next. My initial drafting process, to paraphrase author Shannon Hale, was like shoveling sand into a box so I could later build a castle. Only in the editing stage did I activate convergent thinking and focus on building a meaningful castle out of the sand I had collected (much of which, by the way, had to be thrown away). But when there's just a blank sheet, we need to keep an open mind and not let the castle building predominate the sand collecting.

Beginning with divergent thinking is also important because, at the initial stages of idea formation, it's hard to judge what's useful and what's not. When Benjamin Franklin was watching the first hot-air balloon with humans aboard take off in 1783, someone asked him, "What good is flight?" Franklin purportedly replied, "It is a child who is just born, one cannot say what it will become."[35] Setting the miracle of flight aside, who could fathom in the eighteenth century that balloons would one day be used to distribute a magical technology called the internet to the far reaches of the globe?

Fast forward to the twenty-first century. Within one decade, divergent thinking produced three very different ways of landing on Mars across three missions. The Mars Exploration Rovers launched in 2003 used rovers cocooned in airbags, and the Phoenix mission launched in 2008 used a legged lander.[36] But these landing mechanisms wouldn't work for *Curiosity*, a one-ton rover—more like a Humvee—launched in 2011 with a payload ten times the mass of previous rovers.[37] To land the massive rover gently on the Martian surface, the team strapped an eight-engine jetpack on the rover's back. The jetpack lowered the rover to the surface, separated from the rover, throttled up again, and then crash-landed several hundred yards away from the first spot. The rover's landing system resembled "something Wile E. Coyote might rig up with the ACME Company products," as NASA engineer Adam Steltzner describes it.

Jaime Waydo, who led the design of *Curiosity*'s mobility system, is a fan of far-fetched solutions. "I worry that we are programming people to do the safe thing," she told me. "But safe answers will never change the world."

This belief in expanding what's seemingly possible dates back to Waydo's early schooling. Her math teacher, impressed by Waydo's acumen for math and science, told her that she should think about becoming an engineer. "Isn't engineering something that men do?" Waydo asked him. "When my mom went to college," Waydo explained to me, "she could be a teacher or a psychologist since that's what women did. In her generation, there were clear roles for women in the workforce."

But Waydo's math teacher encouraged her to disregard the historical gender imbalance in engineering and pursue what seemed to her like a gender moonshot. She went on to earn degrees in mechanical and aerospace engineering and, on graduation, took a job at NASA's Jet Propulsion Laboratory to design Martian rovers—joining the ranks of a burgeoning number of women in the previously men-choked corridors of rocket science.

For those tempted to play it safe—by assuming that the light points the only way out of the bottle—Waydo advises keeping the payoff in mind.

Taking a risk on big ideas—using a jetpack to land a Humvee on Mars or building a career that defies stereotypes—is easier when the potential reward is also big. The reward, in the case of *Curiosity*, "is that we have a Humvee driving around on Mars, exploring it, and unlocking the secrets of the solar system," Waydo said. And the reward for Waydo? She helped put three rovers on Mars and later moved on to building self-driving cars—accomplishments that transcend Waydo to enrich every person touched by her skills.

If you're still having trouble activating those divergent-thinking muscles, even with the payoff in mind, the next section will give you a jetpack you can use to boost your own vision.

Shocking the Brain

There was a guy who became famous in the 1970s by lifting some weights. You may have heard of him. You may have seen one or two of his movies. He may have even governed your state.

The biggest obstacle to successful weight training, according to Arnold Schwarzenegger, "is that the body adjusts so quickly." He writes: "Do the same sequence of lifts every day, and even if you keep adding weight, you'll see your muscle growth slow and then stop; the muscles become very efficient at performing the sequence they expect."[38]

Muscles, in other words, have a memory. After sticking to a monotonous routine, they start thinking, *I know exactly what you're going to put me through today. You're going to get on the treadmill and run for thirty minutes, plus or minus three. Every Monday, you'll do bench presses and chin-ups. I'm onto you, and I can handle it.* Schwarzenegger's solution to stagnation was to shock the muscles—to give them exercises of varying types, repetitions, and weights that his muscles hadn't adapted to yet.[39]

Regular makes vulnerable. Irregular makes nimble.

Brains work the same way. Left to its own device, your mind strives for the path of least resistance. Comfortable though it may be, order and predictability get in the way of creativity.[40] We must provoke and shock our minds the same way that Schwarzenegger shocked his muscles.

Neuroplasticity is a real thing. Your neurons, just like your muscles, can rewire and grow through discomfort. As Norman Doidge, a leading expert in neuroplasticity, explains, the brain can "change its own structure and function in response to activity and mental experience."[41] Through reps and sets, thought experiments and moonshot thinking force our minds to rise above our daily trance.

This is why *impossible* was the best compliment one could get from the Nobel-winning physicist Richard Feynman. To Feynman, *impossible* didn't mean unachievable or ridiculous.[42] Rather, it meant "Wow! Here is something amazing that contradicts what we would normally expect to be true. This is worth understanding!" Michio Kaku, the cofounder of string theory, would agree. "What we usually consider impossible are nothing but engineering problems," he says. "There's no law of physics preventing them."[43]

Research supports the link between cognitive contradictions and creativity. When we are exposed to what psychologists call a meaning threat— something that doesn't make sense—the resulting sense of disorientation can prompt us to look for meaning and association elsewhere.[44] As Adam Morgan and Mark Barden write, ideas that appear contradictory "confuse us just enough to start wiring new synapses together."[45] In one study, reading an absurd short story by Franz Kafka, accompanied by equally absurd illustrations, boosted the participants' ability to recognize novel patterns (in other words, connect apples and oranges).[46]

One way to shock your brain and generate wacky ideas is to ask, What would a science-fiction solution look like? Fiction transports us to a reality far different from our own—without the need to ever leave our couch. "Anything that one man can imagine," Jules Verne said, "another man can make real."[47] The thought experiment that gave rise to Project Loon's balloon-powered internet seems as if it came straight out of Verne's *Around the World in Eighty Days*. Verne's other books, including *Twenty Thousand Leagues Under the Sea* and *The Clipper of Clouds*, inspired the creators of the submarine and the helicopter.[48] Robert Goddard, who invented the first liquid-fueled rocket, was transfixed by H. G. Wells's *The War of the Worlds*, a novel about a Martian invasion, and decided to dedicate his life to making spaceflight possible. The science-fiction author Neal Stephenson was one of the first employees of Bezos's Blue Origin. Stephenson was tasked with dreaming up ways of getting to space without conventional rockets (his ideas included using space elevators and lasers that could propel spacecraft).[49]

Science-fiction thinking isn't reserved only for major inventions. Consider a company that produces aircraft parts.[50] Its inspection process was unnecessarily long, primarily because properly inserting a camera into an aircraft part took seven hours. An administrative assistant, inspired by the movie *Minority Report*, posed a thought experiment: "Why can't we send a robotic spider into the part, like the ones in the movie?" The chief technology officer was intrigued. He tested the idea, and it worked spectacularly. This simple fix reduced the inspection time by 85 percent.

Musk credits Asimov's books for spurring his thinking about the future (so much so that SpaceX launched Asimov's Foundation trilogy aboard the *Falcon Heavy* vehicle in February 2018). In the Foundation series, a visionary named Hari Seldon foresees dark ages lurking for humankind and devises a plan to colonize distant planets. "The lesson I drew from that," Musk says, is that humans should "prolong civilization, minimize the probability of a dark age and reduce the length of a dark age if there is one."[51]

People who, like Musk, profess to turn science fiction into fact are often labeled unreasonable. And Musk certainly does his part to boost that image. Every time he opens his mouth, he gives you a reason to doubt him. Aerospace consultant Jim Cantrell, recalling their initial encounters, thought Musk was out of his mind.[52] When Musk first began thinking about a Mars mission, he called Cantrell out of the blue, introduced himself as an internet billionaire, and told Cantrell about his plans to create a "multiplanetary species." Musk offered to fly his private jet to Cantrell's house, but Cantrell said no. "Tell you the truth," Cantrell recalls, "I wanted to meet him in a place where he couldn't bring a weapon." So they met at an airport lounge in Salt Lake City. As wild as Musk's vision sounded, it was too tantalizing. "Okay, Elon," Cantrell said, "let's put a team together and see how much this is going to cost."[53]

Tom Mueller, a founding employee of SpaceX, has often had the same reaction to Musk. "There were times when I thought [Musk] was off his rocker," he says. When the two first met, Mueller was a frustrated rocket scientist at TRW, a large aerospace company that was later acquired by Northrop Grumman. Mueller felt his ideas about engine design were being lost in red tape, so he began designing engines in his own garage.[54] Musk visited Mueller and asked him if the engineer could build a cheap but reliable rocket engine for SpaceX.[55] "How much do you think we can get the cost of an engine down?" Musk asked. Mueller responded, "Oh, probably a factor of three." Musk said, "We need a factor of 10." Mueller thought the answer was pure fantasy. "But in the end," Mueller said, "we're closer to his number!"[56]

To be a universe-denter, you must be unreasonable enough to think you can dent the universe. And unreasonable? That's a label often applied to someone who does something we don't understand. It was the height of unreason to assert that the Earth was round, not flat, or that it revolved around the Sun, not the other way around. When Goddard suggested that rockets could function in the vacuum of space, the *New York Times* ridiculed him. "That professor Goddard, with his 'chair' in Clark College . . . only seems to lack the knowledge ladled out daily in high school," wrote the newspaper in a 1920 editorial. (The paper later issued an apology to Goddard.)

Kennedy's promise of the Moon in less than a decade? Impossible. Marie Curie's attempts to break gender barriers in science? Preposterous. Nikola Tesla's vision of a wireless system for transmitting information? Science fiction.

Often, our moonshots aren't impossible enough. If people want to chuckle at your seeming naivete or call you unreasonable, wear it as a badge of honor. "Most highly successful people have been really right about the future at least once at a time when people thought they were wrong," Sam Altman writes. "If not, they would have faced much more competition."[57] Today's laughingstock is tomorrow's visionary. You'll be the one laughing when you cross the finish line.

Shocking the brain through moonshot thinking doesn't mean we stop considering practicalities. Once we have our wacky ideas, we can collide them with reality by switching from divergent to convergent thinking— from idealism to pragmatism. In the next two sections, we'll learn from two companies that have institutionalized this mindset.

The Business of Moonshots

Designing moonshots for X wasn't on Obi Felten's agenda when she got a phone call from Astro Teller, X's head. Felten is the modern-day Renaissance woman, a polymath as comfortable talking with engineers about hardware as she is building a marketing plan. She grew up in Berlin and saw the wall come down. She then went to Oxford, picking up degrees in philosophy and psychology. She later joined Google as the director of consumer marketing for Europe, Middle East, and Africa.[58] While she was at the top of her marketing game, a phone call from Teller changed everything.

On the call, Teller walked Felten through the audacious projects that X was incubating—including self-driving cars and balloon-powered internet. She responded with questions that Teller hadn't heard before: Is what you're doing legal? Have you talked to any governments and regulators about it? Will you collaborate with other companies? Do you have a business plan?[59]

Teller had no answers. "Oh, no one's really thinking about any of these problems," he replied. "It's all engineers and scientists, and we're just thinking about how to make the balloons fly."

So Felten came on board to think about the practical problems. X may be a moonshot factory, but it's still a factory. It must produce profitable products. "When I came here," Felten explained, "X was this amazing place full of deep, deep, deep geeks, most of whom had never taken a product out into the world."[60]

Pure idealists don't make for great entrepreneurs. Consider Tesla, one of the greatest inventors of all time. "It's a sad, sad story," Larry Page, Google's cofounder explains. "He couldn't commercialize anything, he could barely fund his own research."[61] Although Tesla—whom Edison pejoratively called a "poet of science"—left behind a legacy of three hundred patents, he died penniless in a New York hotel.[62] Reflecting on this story, Page says, "You've got to actually get [your invention] into the world; you've got to produce, [and] make money doing it."

To get X's inventions into the real world, Felten was named "head of getting moonshots ready for the real world" (yes, that's her real title). During her first year at X, she led the company's marketing efforts, built out the legal and government relations teams, and wrote Loon's first business plan.[63]

When X first starts spinning ideas for moonshots, divergent thinking predominates. "At the very early stages of idea formation," Felten told me, "there's tremendous value to science-fiction thinking. If it doesn't break the laws of physics, the idea is potentially fair game."[64]

These ideas are cultivated by a multidisciplinary team of polymaths ideally situated for combinatory play. "The best ideas come from great teams," Felten says, "not great men."[65] X takes cognitive diversity to a new level. The company's ranks include firefighters and seamstresses, concert pianists and diplomats, politicians and journalists. You might find an aerospace engineer working with a fashion designer or a special operations veteran throwing ideas around with a laser expert.[66]

X's goal is to make moonshot thinking the new norm. To this end, the company aims to consistently shock the collective mental muscles of the team. One such exercise is a "bad-idea brainstorm." This might strike you as odd—why waste time with bad ideas?—but X is onto something. "You can't get to the good ideas without spending a lot of time warming up your creativity with a bunch of bad ones," Teller explains.[67] "A terrible idea is often the cousin of a good idea, and a great one is the neighbor of that."

As ideas for potential moonshots begin to move down the funnel, divergent thinking morphs into convergent thinking. The first stage, where wacky ideas collide with reality, is called rapid evaluation. The job of the rapid evaluation team is not only to generate outlandish ideas but also to kill those ideas before X begins to pour money and resources into them. At this stage, as X's Phil Watson explains, "The first thing we're asking is: Is this idea achievable with technology that will be available in the near term, and is it addressing the right part of a real problem?"[68] Only a few of these ideas—those that strike the "right balance of audacity and achievability"— survive rapid evaluation to move on to the next phase.[69]

When the idea of balloon-powered internet entered the rapid evaluation phase, its prospects looked grim. "I thought I was going to be able to prove it impossible really quickly," recounts X's Cliff Biffle. "But I totally failed. It was really annoying."[70] As radical as the solution was, Biffle realized it was actually doable.

If an idea survives rapid evaluation, different teams led by Felten and others pick up the baton. These teams take the science-fiction technologies and lay the foundations for turning them into profitable businesses that solve real-world problems. "Within a year," Felten explained, "we either de-risk the project to a point where we are ready to grow it—or we kill it."[71]

During this de-risking process, Project Loon's balloon-powered internet proved its worth. The preliminary tests—officially called the Icarus tests for the team's audacious goal to fly high—looked promising.[72] But there was a problem. Just as Icarus's wings melted at high altitudes, the balloons would deflate after only five days—far short of their expected continuous circulation of one hundred days. The balloons seemed to be suffering the same type of leakage problem that causes your everyday balloon to deflate into a sad shape the day after a birthday party. The team—which, at this point, was named Daedalus, after Icarus's craftsman father—worked on a fix. They compared apples and oranges, looking for ideas from other industries where leaks also matter. For example, they examined how the food industry makes snack chip bags and sausage casings.[73] They eventually solved the problem and survived all other attempts by Xers to prove the project impossible.

Projects like Loon that survive this rigorous de-risking process graduate from X—with employees getting actual diplomas—and become their own independent companies. X's graduates include businesses that produce self-driving cars, autonomous drones, and contact lenses that measure glucose levels. These ideas all seemed like science fiction—until X struck the right balance between idealism and pragmatism, making them a reality.

At a different company, SpaceX, two leaders represent these two perspectives of idealism and pragmatism. Musk, with his moonshots broadcast liberally from his Twitter account, is the front-facing idealist, the lead singer of the band. Yet someone else behind the scenes has the extraordinarily difficult job of taking Musk's wacky ideas and turning them into actionable businesses.

Her name is Gwynne Shotwell. She's the no-nonsense president and chief operating officer of SpaceX. Shotwell decided to become an engineer when, as a teenager, she attended a Society of Women Engineers event.[74] During one panel discussion, Shotwell was blown away by a mechanical engineer who owned a company that developed environmentally friendly construction materials. The speaker ended up blazing the engineering trail for her.

Now, more than three decades later, Shotwell is at the top of the engineering game, responsible for the day-to-day operations of SpaceX. Among other things, she serves as "the bridge between Elon and the staff," SpaceX's Hans Koenigsmann says.[75] "Elon says let's go to Mars and she says, 'OK, what do we need to actually get to Mars?'" To finance the company's unconventional dream of colonizing Mars, Shotwell travels the globe, pursuing conventional opportunities for taking commercial payloads into orbit. While SpaceX was still in its infancy, she managed to win contracts worth billions of dollars from satellite operators. These contracts continue to pay the bills as SpaceX works toward its moonshot of taking humans to Mars.

But another important question remains: Even if we manage to get to Mars, how will we settle there? Among other things, our Martian pioneers will have to mine raw materials and ice or even build underground tunnels and habitats to shield themselves from long-term radiation.[76]

To perfect tunneling on Mars, we first have to perfect it on Earth. That, in turn, will require the right type of boring technology from the right type of boring company.

A Boring Company

The traffic in Los Angeles is notoriously bad. Depending on the time of day, you might sit for hours in traffic seriously contemplating whether the rest of your life will be spent on the 405 freeway.[77]

If you're a typical city planner tasked with unclogging LA's arteries, the questions are obvious. How do we encourage people to use bikes or public transportation? How do we build more roads? How do we create a carpool lane to decrease rush-hour traffic?

But these questions won't solve the problem. At most, they'll yield incremental improvement. Examine them closely, and you'll find a lack of first-principles thinking. They're all operating with an implied assumption: Traffic is a two-dimensional problem that requires a two-dimensional solution.

Instead of remaining within two dimensions, the Boring Company (yes, that's its real name) posed a thought experiment: *What if we considered a third dimension and went above the ground or under?* In practice, this means either flying cars or driving through underground tunnels.

If you've seen the movie *Back to the Future* as many times as I have, flying cars would seem to be the obvious science-fiction choice ("Roads? Where we're going, we don't need roads!").[78]

But as glamorous as they sound, flying cars have drawbacks. They generate a lot of noise, can be hampered by weather conditions, and can induce anxiety among the earthbound pedestrians about a collision between a flying car and a human head.

In contrast, underground tunnels are weatherproof and invisible to pedestrians above the surface. If you build the tunnels at sufficient depth, their construction and operation generate negligible noise discernible from the surface. Contrary to popular belief, tunnels are among the safest places to be when an earthquake hits. They shield their occupants from falling debris, which can cause great damage during an earthquake. And unlike surface structures, tunnels move with the ground as it shakes. What's more, with an underground tunnel, you can drive from Westwood, California, to the Los Angeles International Airport—a distance of about ten miles—in less than six minutes, instead of sixty minutes in rush-hour traffic.

But here's the problem: It's extremely expensive to dig tunnels—to the tune of hundreds of millions of dollars per mile.[79] This constraint alone can make the project financially prohibitive.

Let's pause here for a moment. We started with divergent thinking (How do we create a three-dimensional solution to traffic congestion?) and allowed ourselves to explore this fantasy without thinking about practical constraints. Now, we'll switch to convergent thinking and take on the financial elephant in the room.

To make underground tunnels affordable, the cost of building a tunnel has to be reduced *tenfold*, which in turn requires tunnel boring machines to become much more efficient. These machines are currently fourteen times slower than a snail—in large part because tunneling technology hasn't improved all that much for the past fifty years. The Boring Company has several ideas for defeating the snail: increase the power output of the machines, improve operations efficiency to reduce downtime, and eliminate human operators by automating the machines. The company also plans to recycle the excavated dirt to build the necessary tunnel structures—which would save money and cut down on concrete use, reducing environmental impact.

In 2018, the city of Chicago selected the Boring Company for exclusive negotiations to build an eighteen-mile tunnel between O'Hare International Airport and downtown Chicago.[80] If the tunnel is built, the trip is expected to take twelve minutes—three to four times faster than the existing transportation methods and at half the typical price of a taxi ride. Las Vegas later followed suit by awarding the company a contract to build a tunnel under its convention center.[81]

Time will tell whether the Boring Company will win its daring dash against the snail. The company's projects are fraught with numerous engineering challenges and potential complications from treacherous geological conditions. But the projects don't have to work. Even if they fail, they're likely to produce improvements in an industry that has stagnated for decades. They will take what was boring and make it exciting.

....

STARRY-EYED DREAMERS AREN'T necessarily known for their follow-through. It's one thing to promise the Moon on a PowerPoint presentation, but it's something else to execute on it. "As for the future," Antoine de Saint Exupéry once wrote, "your task is not to foresee, but to enable it."[82] No matter how creative your moonshot, you'll eventually need to channel your inner Shotwell to ground your vision and figure out how to get there. And getting to the future often requires moving back from it—by using a little-known strategy called backcasting.

Back to the Future

For most of us, planning for the future means forecasting. In our businesses, we review the current supply and demand for widgets and extrapolate them into the future. In our personal lives, we let our current skill set drive our vision for who we might become.

But forecasting, by definition, doesn't start with first principles. With forecasting, we look in the rearview mirror and at the raw materials in front of us, rather than the possibilities ahead. When we forecast, we ask, "What can we do with what we have?" Often, the status quo itself is part of the problem. Forecasting takes all our problematic assumptions and biases and propels them into the future. In so doing, it artificially restricts our vision of what is feasible, given the current circumstances.

Backcasting flips the script. Rather than forecasting the future, backcasting aims to determine how an imagined future can be attained. "The best way to predict the future," Alan Kay says, "is to invent it."[83] Instead of letting our resources drive our vision, backcasting lets our vision drive the resources.

When we backcast, we take our bold ambition and introduce actionable steps. We visualize our ideal job and sketch out a roadmap to get there. We picture the perfect product and ask what it takes to build it. Only when you face the real prospect of sketching a blueprint for success—now, not later—will you be forced to separate fact from fiction.

Backcasting enabled humankind's first actual moonshot. NASA began with the result of landing humans on the Moon and worked backward to determine the steps necessary to get there: Get a rocket off the ground first, then put a person in orbit around Earth, then do a spacewalk, then rendezvous and dock with a target vehicle in Earth orbit, and then send a manned spacecraft to the Moon to circle around it and come back. Only after these progressive steps in the roadmap were completed was a Moon landing attempted.

Amazon takes a similar backward perspective on its products.[84] Amazonians write internal press releases for products that don't yet exist. Each press release functions as a thought experiment—the initial vision of a breakthrough idea. The document describes the "customer problem, how current solutions (internal or external) fail, and how the new product will blow away existing solutions." The press release is then presented to the company with the same enthusiasm that accompanies the public launch of a finished product. "We only fund things that we can articulate crisply," explained Amazon's Jeff Wilke.

The articulation is so crisp that the press releases include a six-page list of hypothetical frequently asked questions from customers. This exercise forces the team of experts at Amazon to put themselves in the position of nonexperts and view the product from their perspective. It requires them to ask "stupid" questions and come up with answers—even before the product is built.

Through backcasting, Amazon can inexpensively evaluate whether ideas are worth pursuing. "Iterating on a press release," Amazon's Ian McAllister explains, "is a lot less expensive than iterating on the product itself (and quicker!)." Backcasting also allows Amazon to focus on its ultimate goal of customer satisfaction. In writing press releases, Amazon doesn't work backward from a finished product. Instead, it works backward from a happy customer. To that end, the press release includes a testimonial from a hypothetical customer gushing about the product. But the press release isn't an exercise in self-deception that assumes the product will wow all customers. In writing their press releases, Amazon employees also ask, "What will customers be most disappointed about in version one of the offering?"

The press release, once written, isn't shelved. It guides the team throughout the entire development process. At each stage, the team asks, "Are we building what's in the release?" If the answer is no, it's time to pause and reflect. Any significant deviation from the initial trajectory may mean that a course correction is necessary.

Yet, it's equally important not to treat the press release as a bible. As entrepreneur and author Derek Sivers writes, "Detailed dreams blind you to new means."[85] The initial specifics in your press release may have a short half-life as the world around you changes. These outdated details shouldn't smother the overall vision. In other words, don't stay the course just for the sake of staying the course.

In getting us to take a hard look at the path to a destination, backcasting can also provide a sobering reality check. We often fall in love with a destination, but not the path. We don't want to climb a mountain. We want to have climbed a mountain. We don't want to write a book. We want to have written one.

Backcasting reorients you toward the path. If you want to climb a mountain, you'll imagine training with your backpack on, hiking at high altitudes to get used to the low-oxygen environment, climbing stairs to build up muscles, and running to improve endurance. If you want to write a book, you'll imagine sitting in front of your computer every single day for two years putting one awkward word after the next, writing one ghastly draft chapter after another, polishing, tweaking, and retweaking—even if you don't feel like it—with no recognition or accolades.

If you go through this exercise, and the idea sounds like torture, then stop. If any of this seems strangely fun to you—as writing does to me—then by all means, go for it. With this reorientation, you also condition yourself to derive intrinsic value from the process rather than chasing elusive outcomes.

Once you have the roadmap ready, it's time to apply the monkey-first strategy.

Monkey First

You've just been put in charge of a particularly audacious project at work. Your boss says you have to get a monkey to stand on a pedestal and train it to recite passages from Shakespeare. How do you begin?

If you're like most people, you begin with building a pedestal. At some point, "the boss is going to pop by and ask for a status update," as Teller explains, "and you want to be able to show off something other than a long list of reasons why teaching a monkey to talk is really, *really* hard." You would rather have the boss give you a pat on the back and say, "Hey, nice pedestal, great job!"[86] So you build the pedestal and wait for a Shakespeare-reciting monkey to magically materialize.

But here's the problem: Building the pedestal is the easiest part. "You can always build the pedestal," Teller says. "All of the risk and the learning comes from the extremely hard work of first training the monkey."[87] If the project has an Achilles heel—if the monkey can't be trained to talk, let alone recite Shakespeare—you want to know that up front.

What's more, the more time you spend building the pedestal, the harder it becomes to walk away from moonshots that shouldn't be pursued. This is called the sunk-cost fallacy. Humans are irrationally attached to their investments. The more we invest time, effort, or money, the harder it becomes to change course. We continue to read a terrible book because we already spent an hour reading the first few chapters or pursue a dysfunctional relationship because it has dragged on for eight months.

To counter the sunk-cost fallacy, put the monkey first—tackle the hardest part of the moonshot up front. Beginning with the monkey ensures that your moonshot has a good chance of becoming viable *before* you've poured massive amounts of resources into a project.

The monkey-first attitude requires developing a set of "kill metrics," as X calls them, a set of go/no-go criteria for determining when to press ahead and when to quit.[88] The criteria must be defined at the outset—when you're relatively clearheaded—before your emotional and financial investments might trigger the sunk-cost fallacy and cloud your judgment.

This approach shut down a project called Foghorn at X.[89] The venture seemed promising at first. A member of X had read a scientific paper about taking carbon dioxide out of seawater and turning the carbon dioxide into affordable, liquid fuel with the potential to replace gasoline. The technology sounded like something out of a sci-fi movie, so X—true to its form—took it on.

Before they began morphing fiction into fact, the members of Team Foghorn set a kill metric. At the time, gasoline was eight dollars per gallon in the most expensive markets. The team aimed, within five years, to produce the equivalent of one gallon of gasoline at five dollars, leaving room for a profit margin and other business expenses.

It turned out that the technology was the pedestal. The team found that it was relatively easy to turn seawater into fuel. But the monkey was the cost. The process was expensive, particularly in the face of declining gasoline prices. When the team members understood that the project couldn't survive the kill criteria, they decided to shut it down. As the project lead Kathy Hannun explained, although the decision was painful, "the strong techno-economic model that we developed at the outset of the investigation made it obvious that was the right thing to do."

There's far more certainty in building a pedestal than in getting a monkey to talk. We don't know how to train a monkey, but we know how to build pedestals, so we build them. In our lives, we spend our time doing what we know best—writing emails, attending endless meetings—instead of tackling the hardest part of a project.

And building pedestals isn't completely unjustified. After all, the project requires the monkey to stand on a pedestal. Crafting the pedestal gives us the satisfaction of doing something about the problem and getting some sense of progress—while postponing the inevitable. All this churn feels productive, but it's not. We've built a beautiful pedestal, but the monkey still isn't talking.

Here's the thing: What's easy often isn't important, and what's important often isn't easy.

In the end, we have a choice. We can keep building pedestals and wait for a magical monkey to show up reciting Shakespeare (spoiler: there are no magical monkeys). Or we can focus on the important instead of the easy, and try to teach that monkey to talk, one syllable at a time.

....

THERE'S A SCENE at the beginning of the movie *Apollo 13*. Jim Lovell, the backup commander for the Apollo 11 mission, watches with admiration as Armstrong and Aldrin take their first steps on the lunar surface. "It's not a miracle," Lovell says. "We just decided to go."

This isn't unbounded optimism—an attitude that says once we dream big, the *Eagle* will magically materialize on Tranquility Base. Instead, it's a combination of optimism and pragmatism—the sheer audacity that combines starry-eyed dreams with a step-by-step blueprint for turning the seemingly unreasonable into reality. "The reasonable man adapts himself to the world," George Bernard Shaw famously said, but "the unreasonable one persists in trying to adapt the world to himself. Therefore all progress depends on the unreasonable man."[90]

That's my moonshot for you: Be more unreasonable. Breakthroughs, after all, are reasonable only in hindsight. "The day before a major breakthrough, it is just a crazy idea," says aerospace engineer Burt Rutan, who designed the first privately funded spacecraft to reach space.[91] If we restrict ourselves to what's possible given what we have, we'll never reach escape velocity and create a future worth getting excited about.

In the end, all moonshots are impossible.

Until you decide to go.

Visit **ozanvarol.com/rocket** to find worksheets, challenges, and exercises to help you implement the strategies discussed in this chapter.

STAGE TWO
ACCELERATE

In this second stage of the book, you'll learn how to propel the ideas you devised in the first stage. You'll discover how to reframe questions to generate better answers; why proving yourself wrong is the path to finding what's right; and how to test and experiment like a rocket scientist to make sure your moonshot has the best shot at landing.

5

WHAT IF WE SENT TWO ROVERS INSTEAD OF ONE?

How to Reframe Questions to Generate Better Answers

A problem well defined is a problem half solved.

—UNKNOWN

TO LAND ON Mars is to execute a perfect cosmic choreography.[1] "If any one thing doesn't work just right, it's game over," NASA engineer Tom Rivellini explained.[2]

For one thing, Mars is a rapidly moving target. Depending on its alignment with Earth, the red planet is between 35 million and 250 million miles away, orbiting the Sun at over 50,000 miles per hour.[3] Landing on a specific site, at a specific time, requires nothing short of an interplanetary hole in one.

But the most dangerous part of the interplanetary journey isn't the six months it typically takes a spacecraft to travel from Earth to Mars when the two planets are closest to each other. Rather, it's the six minutes of terror at the very end of that journey, when the spacecraft enters, descends, and (hopefully) lands on the surface.

During its journey, a typical Mars-bound lander rests inside a two-part aeroshell—a cocoon of sorts—with a heat shield in the front and a back shell on the opposite side. When the spacecraft touches the Martian atmosphere, it is barreling through space at more than sixteen times the speed of sound. In about six minutes, it must bleed off its 12,000-miles-per-hour velocity to land safely on the surface. As the spacecraft tears through the atmosphere, the temperatures outside climb up to over 2,600°F (or roughly 1,400°C). The heat shield protects the spacecraft from bursting into flames as the atmospheric friction slows it down to about 1,000 miles per hour.

That's still really fast. At about six miles above the surface, the spacecraft deploys a supersonic parachute and jettisons the heat shield. But the parachute itself isn't enough to slow down the spacecraft. The Martian atmosphere is thin—its density is less than 1 percent of Earth's atmosphere—and parachutes work by creating drag with air molecules. The fewer molecules, the less drag. As a result, a parachute can bring down the spacecraft's speed to only about 200 miles per hour. We need something else to reduce that velocity so the spacecraft doesn't strike the surface at race-car speed.

In 1999, when I started working on the operations team for what would later become the Mars Exploration Rovers mission, that "something else" was a three-legged lander with rocket motors. After a parachute had reduced its speed, the lander would deploy the three shock-absorbing legs that had been tightly stowed away during the journey. The lander would then fire up its rockets, and using a radar, navigate down to the surface for a soft, steady touchdown on its three legs.

That was the theory. But there was a practical problem. The 1999 Mars Polar Lander, which used this landing system, died a swift death. A NASA review board concluded that the Lander had probably plummeted to the surface after a premature shutdown of its rocket motors.

From our perspective, this accident presented a significant challenge. We were planning to use the same landing mechanism as the Mars Polar Lander, and that mechanism had just failed spectacularly. Our mission was grounded.

Initially, we asked the obvious questions: How can we innovate on the flawed design of the Mars Polar Lander? How do we design a better three-legged lander to ensure a smooth landing? But these questions, as we'll discover, weren't the right questions to be asking.

This chapter examines the importance of searching for a better question instead of a better answer. In the first part of this book ("Launch"), you learned how to reason from first principles and ignite your thinking by conducting thought experiments and taking moonshots to generate radical solutions to thorny problems. But often, the question we originally conceived isn't the best one to ask, and the first problem we identified isn't the best one to tackle.

In this chapter, we'll explore how to resist the initial framing of our questions and discover the importance of finding—rather than solving—the right problem. You'll learn the two seemingly simple questions that salvaged the Mars Exploration Rovers mission and the strategy that Amazon used to create its most profitable division. I'll explain what you can learn from a challenge that most Stanford students failed and why expert chess players perform poorly when they see a familiar move on the board. You'll also discover how the same question gave us a breakthrough technology we use every day, revolutionized an Olympic event, and produced a transformative marketing campaign.

The Sentence Before the Verdict

The way that most people solve problems reminds me of a scene from *Alice's Adventures in Wonderland*. In the scene, the Knave of Hearts is on trial for supposedly stealing tarts. After the evidence is presented, the King of Hearts, who's presiding as a judge, says, "Let the jury consider their verdict." The impatient Queen of Hearts interrupts and retorts, "No, no! Sentence first. Verdict afterwards."

In solving problems, we instinctively want to identify answers. Instead of generating cautious hypotheses, we offer bold conclusions. Instead of acknowledging that problems have multiple causes, we stick with the first cause that pops to mind. Doctors assume they have the right diagnosis, which they base on symptoms they have seen in the past. In boardrooms across America, executives, eager to appear decisive, fall over each other to be the first to deliver the correct answer to a perceived problem.

But this approach puts the cart before the horse—or the sentence before the verdict. When we immediately launch into answer mode, we end up chasing the wrong problem. When we rush to identify solutions—when we fall in love with our diagnosis—our initial answer hides better ones lurking in plain sight. When the sentence is announced first, the verdict is always the same: guilty. The difficulty lies, as John Maynard Keynes said, "not in the new ideas, but in escaping from the old ones."[4]

When we're familiar with a problem, and when we think we have the right answer, we stop seeing alternatives. This tendency is known as the Einstellung effect. In German, *einstellung* means "set," and in this context, the term refers to a fixed mental set or attitude. The initial framing of the question—and the initial answer—both stick.

The Einstellung effect is partly a relic of our education system. In schools, we're taught to answer problems, not to reframe them. The problems are handed to—well, more like forced on—students in the form of problem sets. The phrase *problem set* makes this approach clear. The problems have been set (*einstellung*), and the student's job is to solve them—not change or question them. A typical problem declares "*all* of its constraints, all of its given information, comprehensively and in advance," as high-school teacher Dan Meyer explains.[5] The students then take the prepackaged and preapproved problem and plug it into a formula they memorized, which, in turn, spits out the right answer.

This approach is wildly disconnected from reality. In our adult lives, problems often aren't handed to us fully formed. We have to find, define, and redefine them ourselves. But once we find a problem, our educational conditioning kicks in, launching us into answer mode rather than asking whether there's a better problem to solve. Although we pay lip service to the importance of finding the right problem, we double down on the same tactics that have failed us in the past.

Over time, we become a hammer, and every problem looks like a nail. In a survey of 106 senior executives spanning ninety-one corporations in seventeen countries, 85 percent agreed or strongly agreed that their businesses were bad at defining problems and that this weakness, in turn, imposed significant costs.[6] A different study by management scholar Paul Nutt found that business failures happen in part because problems aren't defined properly.[7] For example, when businesses spot an advertising problem, they search for an advertising solution, artificially excluding all other possibilities. In the study, managers considered more than one alternative in less than 20 percent of their decisions. This environment is hostile to innovation. "Preconceived solutions and limited searches for options," Nutt concluded, "are recipes for failure."[8]

Consider another study. Researchers divided expert chess players into two groups and gave them a chess problem to solve.[9] The players were asked to achieve checkmate using the fewest possible moves. For the first group of players, the board contained two solutions: (1) a solution that was familiar to any skilled chess player and would achieve checkmate in five moves and (2) a less familiar, but better, solution that would produce a checkmate in three moves.

Many experts in the first group couldn't find the better solution. The researchers tracked the players' eye movements and found that they spent much of their time tracing the familiar solution on the board. Even when they claimed to be searching for alternatives, the experts literally couldn't keep their eyes away from what they knew. When they could see the familiar solution—the hammer to their nail—their performance was effectively reduced by three standard deviations.

For the second group of players in the study, the researchers changed the setup so that the familiar solution was no longer an available option. Instead, only the optimal solution would achieve checkmate. Without the familiar solution to distract them, the experts in this second group all found the best solution. In the end, the study confirmed a statement attributed to several world chess champions: "When you see a good move, don't make it immediately. Look for a better one."

When the Einstellung effect gets in the way—when we can't see the better move—we can change our definition of the problem by questioning the question.

Questioning the Question

Mark Adler defies all engineer stereotypes. He's charming and charismatic, with a pair of sunglasses often hanging from his neck—a relic of his upbringing in sunny Florida. He laughs often but also has a strong undercurrent of intensity. In his spare time, he flies small airplanes and goes scuba diving. And he talks as fast as he thinks: My interview with him lasted for over an hour, and I squeezed in three questions at best.

When the Mars Polar Lander crashed in 1999, Adler was an engineer at NASA's Jet Propulsion Laboratory. Recall that our Mars mission was canceled because we were planning to use the same three-legged landing system as the one on the Mars Polar Lander. At the time, everyone involved in our mission—except Adler—was suffering from the Einstellung effect. Like the expert chess players, we focused on the familiar solution on the chess board, which, in our case, was the three-legged lander.

But Adler came up with a better problem to solve. When I asked him about his thought process, he told me it was "really, really simple." The way Adler saw it, our problem wasn't the lander. It was gravity. We were preoccupied with the obvious question: "How do we design a better three-legged lander?" Adler stepped back and asked, "How do we defeat gravity and land our rover safely on Mars?" The same force that causes an apple to fall from a tree also causes unhappy meetings between a spacecraft and the Martian surface unless you do something to cushion the fall.

Adler's solution was to abandon the three-legged lander design. Instead, he proposed using giant airbags with our rover cocooned inside a lander. These balloons would inflate shortly before impact with the Martian surface. Cushioned by these big white grapes, our robot geologist would be released from a height of about ten meters, strike the surface, and bounce roughly thirty to forty times before coming to rest.[10]

Yes, the balloons were crude. Yes, they were ugly as hell. But they worked. Airbags had successfully landed the *Pathfinder* spacecraft on Mars in 1997. Adler knew "they could work because they worked before."

Adler took his proposal to JPL's Dan McCleese, chief scientist for Mars exploration, and asked why it wasn't being considered. McCleese said, "It's because there's no champion for that." So Adler decided to become its champion. He pitched the idea to some of the best people at JPL and got them on board. In less than four weeks—a record time for designing a mission—they put together a mission concept using *Pathfinder*'s landing system. The proposal eventually became reality. NASA selected Adler's design largely because it had the highest probability of getting the spacecraft safely to Mars.

"Every answer," Harvard Business School professor Clayton Christensen says, "has a question that retrieves it."[11] The answer is often embedded within the question itself, so the framing of the question becomes crucial to the solution. Charles Darwin would agree. "Looking back," he wrote in a letter to a friend, "I think it was more difficult to see what the problems were than to solve them."[12]

Think of questions as different camera lenses. Put on a wide-angle lens, and you'll capture the entire scene. Put on a zoom lens, and you'll get a close-up shot of a butterfly. "What we observe is not nature itself, but nature exposed to our method of questioning," said Werner Heisenberg, the brains behind the uncertainty principle in quantum mechanics.[13] When we reframe a question—when we change our method of questioning—we have the power to change the answers.

Research supports this conclusion. A meta-analysis of fifty-five years of research on problem finding across numerous disciplines found a significant positive relationship between problem framing and creativity.[14] In one famous study, Jacob Getzels and Mihaly Csikszentmihalyi found that the most creative art students spend more time in the preparation and discovery stage than do their less creative counterparts.[15] Problem finding, according to these researchers, doesn't end with the preparation stage. Even after spending time viewing the problem from different angles, the more creative individuals keep an open mind as they enter the solution stage and stand ready to make changes to their initial definition of the problem.

In our Mars mission, Adler was like the more creative art students, spending more time in formulating the problem and seeing a question that everyone else had missed. But what happened next even Adler couldn't foresee.

The Doppelganger

In many ways, Mars is Earth's sister planet. It's next in line from the Sun. Its seasons as well as its rotational period and the tilt of its axis are similar to ours. Although Mars is cold and desolate now, it was warmer and wetter in the past, with evidence that liquid water flowed across its surface.

These characteristics make Mars one of the few places in the solar system where extraterrestrial life could have existed—even thrived. After the final Apollo mission to the Moon in 1972, Mars naturally appeared to be the next frontier. The Mariner probes, a series of spacecraft launched between 1962 and 1973, had already snapped photos of the red planet from orbit.[16] It was time to get down to the surface. If NASA astronauts could do what Armstrong and Aldrin did—put on a spacesuit and head to Mars with hammers, scoops, and rakes to collect samples—they would have done it. But from NASA's perspective, that option wasn't financially feasible. So NASA did the next best thing: Instead of human geologists, it sent robotic ones.

NASA's first attempt to land on Mars came in 1975 with the launch of the Viking mission. Named after the Nordic explorers, the mission sent two identical space probes to Mars, unimaginatively named *Viking 1* and *Viking 2*.[17] The probes each contained an orbiter designed to analyze the planet from Martian orbit and a lander to study its surface. After the spacecraft arrived at Mars, the orbiters spent some time scouting suitable landing sites. When the landing sites were spotted, the landers detached from the orbiters and descended toward the surface.

The *Viking 1* lander touched down on July 20, 1976—seven years from the date of the *Eagle*'s touchdown on Tranquility Base—and was followed by *Viking 2* in September of that year. Designed to last for ninety days, both landers significantly outlived their warranty. The *Viking 1* lander conducted science for over six years and *Viking 2* for nearly four, beaming tens of thousands of images back to Earth.[18]

Some of these images dotted the entryway of Cornell's Space Sciences Building, where I spent much of my time as an undergraduate. A giant smile would automatically appear on my face as I strolled past them each day before heading up to my fourth-floor work space in the Mars Room. If there was a montage of my college life, the Viking images would feature prominently.

Sometime in 2000, I was busy designing operations scenarios in the Mars Room, simulating what would happen after our rover landed on Mars. This was after Adler's brilliant insight with the airbags had brought us back from the dead. I heard the distinctive sounds of Steve Squyres's boots clicking toward my colleagues and me in the hallway. Squyres, my boss and the principal investigator of our mission, walked into the room and announced that he had just gotten off the phone with Scott Hubbard at NASA headquarters.

When it comes to creating worst-case scenarios, my imagination is particularly vivid. Pessimistic thoughts immediately began clanging in my head. What had gone wrong this time? Were we being scrapped again?

But the news wasn't bad. Hubbard was in charge of fixing NASA's Mars exploration program after the Mars Polar Lander accident. He had just left a meeting with NASA administrator Dan Goldin, who had asked Hubbard to relay a simple question to Squyres.

"Can you build two?" Hubbard had asked Squyres on the phone.

Squyres replied, "Two what?"

Hubbard responded, "Two payloads."

Dumbstruck, Squyres asked, "Why would you want two payloads?"

"For two rovers," Hubbard said.[19]

It was a simple question no one had thought of asking before: Can we send two rovers instead of one? After the Mars Polar Lander crash, we had narrowly focused on the problem with our lander and replaced it with Adler's airbag design. But the risk wasn't isolated to the landing system. Any number of random things could break our spacecraft while traveling nearly forty million miles through outer space and landing on a Martian surface littered with scary-looking rocks while getting whipped by strong winds.

Goldin's solution for this uncertainty was to use a strategy we encountered earlier in the book: introduce a redundancy. Instead of putting all our eggs in one spacecraft's basket and crossing our fingers that nothing bad would happen along the way, we decided to send two rovers instead of one. Even if one failed, the other might make it. What's more, with economies of scale, the cost of the second rover would be pennies on the dollar. After Goldin came up with the idea, Adler and another JPL engineer, Barry Goldstein, were given all of forty-five minutes to estimate how much the second rover would cost. They came up with $665 million for two rovers, which was roughly 50 percent more than the $440 million price tag for one.[20] NASA managed to find the extra cash and gave us the green light.

Just like that, our rover birthed a doppelganger.

This time, NASA decided to get more creative with the naming and held a Name the Rovers contest, allowing schoolchildren to submit essays with suggestions.[21] The winner among ten thousand submissions was Sofi Collins, a third-grader from Arizona, who was born in Siberia and had lived in an orphanage until her adoption by an American family: "It was dark and cold and lonely," she wrote in her essay describing her orphanage. "At night, I looked up at the sparkly sky and felt better. I dreamed I could fly there. In America, I can make all my dreams come true. Thank you for the 'Spirit' and the 'Opportunity.'"

The primary scientific goal of the newly christened *Spirit* and *Opportunity* rovers was to determine whether Mars had once been capable of supporting life. Because water is a crucial ingredient for life as we know it, we wanted to go where water had once gone before. Double the rovers also meant double the science. Two rovers could examine two very different landing sites. If one site turned out to be a flop from a science perspective, the other site might save the day.[22]

For *Opportunity*, we chose Meridiani Planum, a plain near the Martian equator. The area appeared promising because its chemical composition—specifically, the existence of a mineral called hematite—suggested a history of liquid water. What's more, Meridiani Planum is one of the "smoothest, flattest, and least windy places" on the red planet, the Martian equivalent of a giant parking lot.[23] In terms of landing sites, it would be hard to find a safer one.

With *Opportunity* heading to a site rich in chemistry, we picked Gusev, a landing site rich in its topography, for *Spirit*. Located on the opposite side of the planet from Meridiani, Gusev is a giant impact crater with a visible channel. Scientists suspected that the channel had been carved by water at some point in the past and that the crater once held a lake. Gusev was slightly riskier from a landing perspective: It had higher winds and higher rock density than Meridiani had. But with two shots on goal, we could afford to take a bit more risk with one of them.

Spirit was the first to arrive at Mars.[24] After the spacecraft touched the Martian atmosphere, things went as planned. The parachute was deployed. The heat shield was jettisoned. The airbags were inflated, to be followed by lots of bouncing and tumbling on the Martian surface until the lander came to rest. Any lingering doubts about whether Adler's airbag design would work disappeared quickly as the first photos of Mars began to flow in. After years of looking at photos of Gusev taken from orbit, it was surreal to see, for the first time ever, the inside of the crater from the Martian surface in all its high-resolution glory.

But the initial thrill of the landing began to wane as our group started analyzing the images in detail. Yes, we were safely on Mars, and yes, the achievement put us in the distinct minority of missions that had successfully landed there. But aside from the fact that we were looking at Mars, what we were seeing was less than thrilling. The images from the rover looked a lot like those taken by the Viking landers and hung in the Space

Sciences Building as decorations: similar rocks, similar outlook, similar structure—similar everything.

This initial scientific whimper would later turn into a bang when *Spirit* began roving the terrain and arrived at Columbia Hills, a range of peaks three kilometers away from our initial landing site. The peaks were named after the seven astronauts who perished in the Space Shuttle *Columbia* disaster one year before our landing. In those hills, Spirit would eventually go on to find goethite—a mineral that forms only in water, strongly indicating that Mars once had water activity above the surface.

Three weeks later, Spirit's twin, *Opportunity*, touched down on Mars. Meridiani Planum, which was *Opportunity*'s landing site, was like nothing we had ever seen before. Every photo ever taken of the Martian surface has chunks of rock scattered across the surface. But where *Opportunity* landed, there were no rocks. When the rover began to beam its first photos of the landing area to Earth, the mission support team at JPL began laughing, cheering, and crying. The flight director, Chris Lewicki, asked Squyres for a quick science overview of what they were seeing on the screen. But Squyres's throat constricted. He slowly flipped the switch on his headset and said, "Holy smokes. I'm sorry, I'm just, I'm blown away by this."

What they were seeing was an outcrop of bedrock right in front of the rover. Why would something as benign as bedrock leave a scientist speechless? An exposed, layered bedrock is the closest thing there is to time travel. A bedrock is like a history book. It shows us exactly what happened a long, long time ago, on this planet far, far away. Unlike *Spirit*—which had to climb a mountain, literally and figuratively, to find interesting science— *Opportunity* was handed scientific secrets on a silver platter, or in this case, on bedrock. All of *Opportunity*'s big discoveries came within the first six weeks of the mission, thanks to its opportunistic landing site—which was made possible by our decision to send two rovers.

Squyres didn't realize it then, but his comments—including the "holy smokes" bit—were broadcast across the globe. They piqued the interest of a journalist in Seoul, South Korea, writing for the daily newspaper

Munhwa Ilbo. The journalist wrote up the story of *Opportunity*'s historic Mars landing, summed up by the following headline: "The Second Mars Rover Lands, Sees Mysterious Smoke." As another Korean journalist observed, it was fortunate Squyres didn't say *holy cow.*

Like their Viking grandfathers, our rovers were designed to operate for ninety days. But they far outlived the Viking landers. *Spirit* lasted for more than six years until it got stuck on soft soil. It eventually lost communication with Earth after winter arrived and deprived its solar panels of their power source. A formal farewell was held for *Spirit*—complete with toasts and spirited eulogies for a rover that climbed mountains (which it wasn't designed to do) and braved intense dust storms.[25]

Opportunity—or *Oppy*, as we lovingly called it—kept going until June 2018, when a giant dust storm covered the rover's solar panels, starving it of power. NASA officials sent hundreds of commands asking *Oppy* to call home, with no success. In February 2019, *Opportunity* was officially pronounced dead—over fourteen years into its ninety-day expected lifetime, having roamed a record-breaking twenty-eight miles on the red planet.[26]

Holy smokes indeed.

In the end, two questions that reframed the problems ended up producing one of the most successful interplanetary missions of all time: What if we used airbags instead of a three-legged lander? What if we sent two rovers instead of one?

These questions may appear obvious, but they were obvious only in hindsight. How do you do what Adler and Goldin did and see the problem from a perspective others miss? One approach is to distinguish between two concepts—strategy and tactics—that are often conflated. To understand the distinction, let's say goodbye to Mars (for now) and head over to Nepal.

Strategy and Tactics

Babies who are born too soon—before certain key organs develop—are called premature babies (or preemies). Worldwide, roughly one million preemies die of hypothermia each year.[27] Because these babies are born with very little body fat, it is difficult for them to control their body temperature.[28] To them, room temperature can feel like freezing cold water.

In developed countries, the solution is to place the baby in an incubator. The size of a standard crib, an incubator keeps babies warm while their bodies finish developing.[29] The original incubators were fairly simple devices, but bells and whistles were added over time. Incubators now have arm ports for handling the baby, life support devices such as ventilators, and equipment to regulate humidity.[30] The boost in tech also brought a boost in cost. A modern incubator is priced between $20,000 and $40,000, a price that doesn't include the electricity required to make it work. Incubators are consequently hard to find in many developing countries, and the result is preventable deaths.

Four Stanford University graduate students set out to tackle this challenge in 2008 and build cheaper incubators.[31] They were enrolled in a course called Design for Extreme Affordability, where students "learn to design products and services that will change the lives of the world's poorest citizens."[32]

Instead of trying to innovate from the comfort of Silicon Valley, the team decided to take a field trip to Kathmandu, Nepal's capital, to immerse themselves in the practices at a neonatal unit. They wanted to observe how incubators were being used in hospitals so that they could design cheaper equipment to work in local conditions.

But a surprise awaited them. The incubators in hospitals were collecting dust and sitting unused. Part of the problem was technical expertise. Incubators are often difficult to operate. What's more, the overwhelming majority of premature babies in Nepal were born in rural areas. Most of these babies would never make it to a hospital in the first place.

The problem, therefore, wasn't the lack of incubators in hospitals. Rather, it was the lack of accessible infant warmers in rural areas with no access to hospitals or, for that matter, reliable electricity. The traditional solution—send more incubators to hospitals or lower their cost—wouldn't move the needle.

In light of this experience, the Stanford team reframed the problem. Premature babies didn't need incubators. Premature babies needed warmth. Sure, other fancy features on modern incubators, like heart-rate monitors, were helpful, but the most important challenge—the one that would have the highest impact—was keeping the baby warm as its organs developed. The device to provide warmth had to be inexpensive and intuitive so it could be used by an often-illiterate parent in a rural environment without reliable electricity.

The result was the Embrace infant warmer. It's a small, light sleeping bag that wraps around the infant. A pouch of phase-change material— which is an innovative wax—keeps the baby at the right temperature for up to four hours. You can "recharge" the warmer in only a few minutes by putting it in boiling water. And compared with the $20,000 to $40,000 price tag of a traditional incubator, the Embrace costs only $25. By 2019, the cheap and reliable product has embraced hundreds of thousands of premature infants in over twenty countries.

Often, we fall in love with our favorite solution and then define the problem as the absence of that solution. "The problem is, we need a better three-legged lander." "The problem is, we don't have enough incubators." In each case, we pursue technology for the sake of technology. We lose the forest for the trees, the purpose for the method, the function for the form.

This approach mistakes tactics for strategy. Although the terms are often used interchangeably, they refer to different concepts. A strategy is a plan for achieving an objective. Tactics, in contrast, are the actions you take to implement the strategy.

We often lose sight of the strategy, fixate on the tactics and the tools, and become dependent on them. But tools, as author Neil Gaiman reminds us, "can be the subtlest of traps."[33] Just because a hammer is sitting in front of you doesn't mean it's the right tool for the job. Only when you zoom out and determine the broader strategy can you walk away from a flawed tactic.

To find the strategy, ask yourself, What problem is this tactic here to solve? This question requires abandoning the what and the how and focusing on the why. The three-legged lander was a tactic, and landing safely on Mars was the strategy. The incubator was a tactic, and saving premature infants was the strategy. If you're having trouble zooming out, bring outsiders into the conversation. People who don't regularly use hammers are less likely to be distracted by the hammer sitting in front of you.

Once you identify the strategy, it becomes easier to play with different tactics. If you frame the problem more broadly as one of gravity—not as a flawed three-legged lander—airbags can present a better alternative. If you frame the problem more broadly as the risk involved in landing on Mars—not just as a defective lander—sending two rovers instead of one decreases risk and increases reward.

Peter Attia, a physician and a renowned expert on human longevity, is a master at distinguishing between strategy and tactics. I asked him what he does when patients come to him looking for the "right answers." *What diet should I follow? Should I take statins if I have high cholesterol?* "I generally do not let patients fixate on tactics," he told me, "and instead try to refocus them on strategy. When people are looking for the 'right answers,' they are almost always asking tactical questions. By focusing on the strategy, this allows you to be much more malleable with the tactics." For Attia, whether to use a statin is "a tactical question that is in service of the much broader strategy" of delaying death from atherosclerosis.[34]

To teach the difference between strategy and tactics to her students, Tina Seelig, the faculty director of the Stanford Technology Ventures Program, uses what she calls the five-dollar challenge.[35] Students break up into teams, and each team gets five dollars in funding. Their goal is to make as much money as possible within two hours and then give a three-minute presentation to the class about what they achieved.

If you were a student in the class, what would you do?

Typical answers include using the five dollars to buy start-up materials for a makeshift car wash or lemonade stand and buying a lottery ticket. But the teams that follow these typical paths tend to bring up the rear in the class.

The teams that make the most money don't use the five dollars at all. They realize that the five dollars is a distracting, and essentially worthless, resource.

So they ignore it. Instead, they reframe the problem more broadly as "What can we do to make money if we start with absolutely nothing?" One particularly successful team made reservations at popular local restaurants and then sold the reservation times to those who wanted to skip the wait. These students generated an impressive few hundred dollars in just two hours.

But the team that made the most money approached the problem differently. The students understood that both the five-dollar funding and the two-hour period weren't the most valuable assets at their disposal. Rather, the most valuable resource was the three-minute presentation time they had in front of a captive Stanford class. They sold their three-minute slot to a company interested in recruiting Stanford students and walked away with $650.

What is the five-dollar tactic in your own life? How can you ignore it and find the two-hour window? Or even better, how do you find the most valuable three minutes in your arsenal? Once you move from the what to the why—once you frame the problem broadly in terms of what you're trying to do instead of your favored solution—you'll discover other possibilities in the peripheries.

Just as you can reframe questions to generate better answers, you can also reframe objects, products, skills, and other resources to put them to more creative uses. That requires thinking outside the box—in this case, the thumbtack box.

Thinking Outside the Thumbtack Box

What is a barometer for?

If you think the only answer is *measuring pressure*, think again.

Science professor Alexander Calandra—who was an advocate of unorthodox teaching methods—once penned a short story titled "Angels on a Pin."[36] In the story, a colleague asks Calandra to be the arbiter of a dispute between the colleague and a student over a question on a physics exam. The physics teacher believes the student deserves a zero, but the student demands full credit.

Here's the question: "Show how it is possible to determine the height of a tall building with the aid of a barometer." The traditional answer is clear: You take pressure measurements with the barometer at the top of the building and at the bottom and use the difference to compute the height.

But this wasn't the answer the student gave. Instead, the student wrote: "Take the barometer to the top of the building, attach a long rope to it, lower the barometer to the street, and then bring it up, measuring the length of the rope. The length of the rope is the height of the building."

The answer is certainly correct. But it's a deviation from the norm. It's not what the teacher had taught in class—the expected path to the expected outcome. A barometer is supposed to measure pressure, not serve as a makeshift weight for a rope.

The barometer story is a good example of functional fixedness. As psychologist Karl Duncker explains, the concept refers to a "mental block against using an object in a new way that is required to solve a problem." Just as we treat problems and questions as fixed, we do the same with tools. Once we learn that a barometer measures pressure, we blind ourselves to other uses for it. Like the chess players whose eyes kept darting to the familiar solution on the chess board, our minds fixate on the function we know.

Perhaps the most famous example of functional fixedness is the candle problem, designed by Duncker. He devised an experiment where he seated participants at a table adjacent to a wall and gave them a candle, some matches, and a box of thumbtacks. He asked them to figure out a way to attach the candle to the wall so the wax doesn't drip on the table below. Most people tried one of two approaches. They attempted to use thumbtacks to pin the candle to the wall or melt the side of the candle with a match to stick it to the wall.

But neither approach works. These participants failed in part because they focused on the traditional functions of the objects: Thumbtacks are for attaching things. Boxes are for storing things.

The successful participants disregarded the traditional function of the box. Instead, they reframed the box as a platform for the candle to stand on. They then affixed the box to the wall using thumbtacks.

We all encounter variations of the candle problem in our personal and professional lives. And often, we do what the unsuccessful participants did and view the box as a container, not as a platform. So how do we train ourselves to think outside the thumbtack box? How do we see the products or the services we provide from a different perspective? How do we take the skills we have in one field and recognize their value in another?

In a study conducted for the military, Robert Adamson attempted to answer these questions.[37] He replicated Duncker's candle experiment but with a twist: He divided the participants into two groups and slightly modified the setup for each. The second group far outperformed the first. Only 41 percent of the participants in the first group solved the puzzle, compared with 86 percent in the second.

What explained the stark difference in the outcome? For the first group, the three types of materials—the candle, the matches, and the thumbtacks—were placed in three boxes. The first group saw boxes being used as containers and, as a result, suffered from acute functional fixedness. They had a much harder time using the box for anything other than storing objects.

But for the second group, the objects were sitting on the table next to—not inside—the boxes, which were empty. With the objects out of the boxes, the participants could more easily see the boxes as potential stands for the candle. The results were similar to the conclusion of the study involving chess experts. In both cases, performance improved when the familiar solution was removed.

Functional fixedness arises from a set of assumptions we have about what a box or a barometer is supposed to do. We can reduce functional fixedness by taking out Occam's razor—which we explored earlier in the book—and cutting our assumptions about the tool. If you didn't know what you know, what else could you do with it? This can be as simple as blocking its obvious use—dumping the materials out of the box (as Adamson's study did), removing the familiar solution from the chess board, or using the barometer for anything other than measuring pressure.

Combinatory play also helps. You can draw inspiration from how objects are used in other fields. For example, the airbags that landed my group's rovers safely on Mars used the same mechanism that cushions a collision with your steering wheel in a car accident. The same fabric used in astronaut spacesuits is also used by Embrace to make a temperature-controlling swaddle.[38] George de Mestral created Velcro after he saw his pants covered in cockleburs following a walk.[39] He examined the cockleburs under a microscope and discovered a hooklike shape that he then emulated to create the hook-and-loop fastener called Velcro—with one side stiff like the cockleburs and the other side smooth like his pants.

It's also helpful to separate function from form. When we look at an object, we tend to see its function. A barometer, we think, is for measuring pressure. A hammer is for driving in nails. A box is for storing objects. But this natural inertia toward the function also gets in the way of innovation. If we can look past the function to the form, we can discover other ways that the product, service, or technology can be used. For example, if you can view the typical barometer simply as a round object, it can also be used as a weight. If you view a box as a flat platform with sides, it can also be used as a stand.

In one study, participants were divided into two groups and asked to solve eight insight problems—including the candle problem—that required them to overcome functional fixedness.[40] The control group received no training. The other group was taught to use function-free descriptions of objects—for example, instead of saying "a prong of an electrical plug," they were taught to describe the prong as "a thin, rectangular piece of metal." The group that received training solved 67 percent more problems than did the other participants.

The switch from function to form is also helpful in reframing the resources at your disposal. Consider, for example, the development of Amazon Web Services (AWS).[41] As Amazon grew from an online bookstore to an "everything" store, it built up an immense electronic infrastructure, including storage and databases. The company realized that its infrastructure wasn't simply an internal resource. It could also be sold to other companies as a cloud-computing service, to be used for storage, networking, and databases. AWS eventually became a cash cow for Amazon, generating roughly $17 billion in revenue in 2017—more than Amazon's retail division.[42]

Amazon reframed the thumbtack box again with its purchase of Whole Foods Market. The purchase befuddled many observers. Why was the internet giant buying up a struggling brick-and-mortar grocery store chain? One answer was based on the reframing of the physical Whole Foods Market stores. Instead of seeing them simply as grocery stores, Amazon reframed them as distribution centers located in densely populated urban hubs. These centers could enable fast delivery of products to Amazon Prime customers.[43]

In both cases, Amazon looked beyond the function to the form. The function of Whole Foods stores was to sell groceries, but the stores took the form of a massive real estate footprint with storage and refrigeration that could be repurposed for distribution. The function of Amazon's computing infrastructure was for internal support, but its form—a massive data center—could provide a highly profitable service to companies such as Netflix and Airbnb.

If you're having a hard time switching from function to form and seeing the thumbtack box as a candle platform, there's another approach you can try: Reverse the box.

What If We Did the Reverse?

On Friday, October 4, 1957, the Soviet Union launched *Sputnik*, the first artificial satellite in Earth orbit.[44] Russian for "fellow traveler," *Sputnik* orbited the Earth roughly every ninety-eight minutes. If you doubted that humankind had created its own moon, you could walk outside with a pair of binoculars after sunset and see it flying overhead.

You could not only see *Sputnik*, but hear it as well. At the time, two young physicists by the name of William Guier and George Weiffenbach were working in the Johns Hopkins Applied Physics Laboratory in Maryland.[45] They were curious whether the microwave signals emitted by *Sputnik* could be received on Earth. In a matter of hours, Guier and Weiffenbach had locked in on a series of signals emanating from the satellite.

Beep. Beep. Beep.

This easily detectable signature was no oversight on the Soviets' part. Masters at propaganda, the Soviets had intentionally engineered the *Sputnik* to broadcast a signal that could easily be picked up by anyone on Earth with a shortwave radio.

Beep. Beep. Beep.

As Guier and Weiffenbach listened to the Red broadcast, they realized they could use the signal to calculate *Sputnik*'s speed and trajectory. Just as the siren of an ambulance buzzing past you decreases in pitch, the beeps from *Sputnik* changed as the satellite moved away from the scientists' location. Using this phenomenon—called the Doppler effect—the two men plotted the entire trajectory of *Sputnik*.

Sputnik's launch inspired awe but also whipped Americans into a frenzy. "If the Russians can deliver a 184-pound 'Moon' into a predetermined pattern 560 miles out in space," wrote an editorial in the *Chicago Daily News*, "the day is not far distant when they could deliver a death-dealing warhead onto a predetermined target almost anywhere on the Earth's surface."[46]

Frank McClure was also thunderstruck by *Sputnik*, but for a different reason. McClure was then the deputy director at the Applied Physics Laboratory. He called Guier and Weiffenbach into his office and asked them a simple question: "Can you guys do the reverse?" If the two men could calculate the unknown trajectory of a satellite from a known location on Earth, could they find an unknown location on Earth using the known location of a satellite?

This question may sound like a theoretical riddle, but McClure had a very practical application in mind. At the time, the military was developing nuclear missiles capable of being launched from submarines. But there was a problem. To strike a precise location with a nuclear missile, the military had to know the precise location of the launch site. In the case of nuclear submarines swimming through the Pacific Ocean, their precise location was unknown. Hence the question: Can you discover the unknown location of our submarines through the known location of a satellite we'll launch into space?

The answer was a resounding yes. It took only three years after the launch of *Sputnik* for the United States to implement this thought experiment and launch five satellites into orbit to guide its nuclear submarines. Although it was called the Transit system at the time, its name was changed in the 1980s to something that has become an everyday term: the global positioning system, or GPS.

McClure's approach illustrates a powerful way of reframing questions: taking an idea and flipping it on its head. This method dates back at least to the nineteenth century, when the German mathematician Carl Jacobi introduced the idea with a powerful maxim: "Invert, always invert" (*Man muss immer umkehren*).[47]

Michael Faraday applied this principle to generate one of the greatest scientific discoveries of all time. In 1820, Hans Christian Ørsted—who coined the term *thought experiment*—discovered the connection between electricity and magnetism. He noticed that a compass needle deflected when a wire carrying an electric current passed over it.

Later, Faraday came along and reversed Ørsted's experiment. Instead of passing a wire with electric current over a magnet, he passed a magnet around a coil of wire. This generated an electrical current that grew bigger the faster he spun the magnet. Faraday's reversal experiment gave way to modern hydroelectric and nuclear power plants, both of which use a magnetic turbine that generates electricity by turning a wire around.[48]

Across disciplines, in biology, Darwin adopted the same reversal mantra.[49] While other field biologists looked for differences between species,

Darwin searched for similarities. He compared, for example, the wing of a bird with the hand of a human. Exploring the similarities between otherwise vastly different species eventually culminated in the theory of evolution.

The power of inversion extends far beyond science. To cite a business example, the clothing company Patagonia reversed an industry best practice in a 2011 advertising campaign.[50] The company asked, "Instead of telling people to buy from us, what if we told them *not* to buy from us?" The result was a full-page ad in the *New York Times* that ran on Black Friday—the Friday after US Thanksgiving, when Americans flock to stores to take advantage of deep discounts for the holiday shopping season. The ad featured a Patagonia jacket with the headline "Don't buy this jacket." With this ad, Patagonia became "the only retailer in the country asking people to buy less on Black Friday."[51] The ad worked, partly because it supported Patagonia's mission of reducing consumerism and lowering environmental impact. But the contrarian ad also ended up helping the company's bottom line by attracting customers who shared the same mindset.

In athletics, reversing conventional wisdom landed Dick Fosbury the Olympic gold medal.[52] At the time, if you had met Fosbury in person, you wouldn't have thought he was an athlete. He was awkward, scrawny, and tall, with a significant acne problem that he couldn't seem to shake. When Fosbury was training to be a high jumper, athletes would use a technique called the straddle method, where they would jump face down over the bar. At the time, the straddle method was considered beyond improvement. There was no need to experiment or come up with something new.

But the straddle method never worked for Fosbury. As a high-school sophomore, he was performing at a junior high level. On a bus ride to a track meet, Fosbury decided to do something about his mediocrity. The rules allowed athletes to clear the bar any way they wanted as long as they jumped off one foot. The straddle method was a mere tactic. But clearing the bar was the strategy. So instead of jumping face down to the bar, Fosbury did the reverse. He jumped backward.

His approach initially invited ridicule. A newspaper called him "The World's Laziest High Jumper."[53] Many fans laughed at him as he cleared the bar like a fish flopping in a boat.

The laughs eventually turned into cheers as Fosbury proved his critics wrong and took home the gold medal at the 1968 Summer Olympics—by doing the exact opposite of what everyone else was doing. The Fosbury flop, as it has come to be known, is now the standard method used at Olympic high jump events. Fosbury came home to a ticker-tape parade and appeared live on the *Tonight Show*, where he taught Johnny Carson how to perform the Fosbury flop.

The serial entrepreneur Rod Drury calls this approach the "George Costanza theory of management."[54] In an episode of *Seinfeld*, Costanza sets out to improve his life by doing the exact opposite of what he had done before. Drury, who founded and led the accounting software company Xero, would outsmart his far bigger competitors by asking himself, "What is the exact opposite of what an incumbent would expect us to do?" Asking this question in 2005, Drury went all in on using a cloud-based platform when his competitors were all still stuck on desktop applications.[55]

Drury knows a secret missed by many business leaders: The low-hanging fruit has already been picked. You can't beat a stronger competitor by copying them. But you can beat them by doing the opposite of what they're doing.

Instead of adopting a common best practice or the industry standard, reframe the question by asking, "What if I did the reverse?" Even if you don't execute, the simple process of thinking through the opposite will make you question your assumptions and jolt you out of your current perspective.

....

THE NEXT TIME you're tempted to engage in problem solving, try problem finding instead. Ask yourself, Am I asking the right question? If I changed my perspective, how would the problem change? How can I frame the question in terms of strategy, instead of tactics? How do I flip the thumb-tack box and view this resource in terms of its form, not its function? What if we did the reverse?

Breakthroughs, contrary to popular wisdom, don't begin with a smart answer.

They begin with a smart question.

Visit **ozanvarol.com/rocket** to find worksheets, challenges, and exercises to help you implement the strategies discussed in this chapter.

6

THE POWER OF FLIP-FLOPPING

How to Spot the Truth and Make Smarter Decisions

It is a capital mistake to theorize before one has data.
Insensibly one begins to twist facts to suit theories,
instead of theories to suit facts.

—SHERLOCK HOLMES

MARS IS A master of deception.[1] Since the beginning of humankind, the red planet has been staring at us as one of the brightest lights in the night sky. With its red hue, the planet might appear warm, cozy, and even welcoming to the unsuspecting observer.

But it's not. Mars is a hostile place—not just because the average surface temperature is −63°C (−81°F), not just because it's drier than the driest deserts on Earth, and not just because it kicks up intense dust storms that span continent-sized areas.[2]

Mars is hostile to us because it hosts the biggest graveyard of human spacecraft.

When I started working on the operations team for the Mars Exploration Rovers project, two out of every three Mars missions had failed. I learned quickly that the red planet wouldn't be rolling out any red carpets for us. Upon entry into the Martian atmosphere, we would be greeted by what's been called the "galactic ghoul," a fictitious Martian monster that feeds on human spacecraft.

On September 23, 1999, the Mars Climate Orbiter became the latest victim of the galactic ghoul. The orbiter was designed to be the first spacecraft to study another planet's weather from its orbit. On the evening the orbiter arrived at Mars, I huddled with the other members of the Mars Exploration Rovers team at Cornell to watch NASA TV while we held our breath. This wasn't our baby, but we had a lot riding on the orbiter's success. The vehicle would serve as our primary radio relay after we landed on Mars. It would communicate our commands to the rovers on the surface and send their responses back to us. It was our walkie-talkie.

The orbiter arrived at Mars as expected. The next step was an orbital insertion burn: The navigation team fired the orbiter's main engine to slow it down and drop it into an orbit around the red planet. As the spacecraft passed behind Mars, its radio signal was blocked by the planet and disappeared as scheduled. We waited, along with the engineers in mission control, for the signal to reappear as the spacecraft sailed back into sight.

But the signal didn't reappear. As the clock continued to tick with no sign of the orbiter, the tenor in the room changed unsettlingly fast. We had just lost our walkie-talkie.

There are no obituaries written for spacecraft devoured by the galactic ghoul. But if there were, the obituary for the Mars Climate Orbiter would read: "A perfectly healthy spacecraft, operated by some of the smartest rocket scientists in the world, was flown straight into the Martian atmosphere, where it died a horrific death."

If your goal is to put a spacecraft in orbit around Mars, you must keep the spacecraft safely above the atmosphere. At low altitudes, the atmosphere becomes hostile. The spacecraft can burn up after grinding too hard into the atmosphere or skip across it and bounce into the endless abyss of space. The orbiter was programmed to enter into an orbit at a safe 150 kilometers above the surface. But instead, it entered Mars at an altitude of just 57 kilometers—deep within the atmosphere.

A NASA press release attributed the nearly 100-kilometer gap to a "suspected navigation error."[3] But in less than a week, it became clear that the "navigation error" was NASA's understatement of the decade. The $193 million spacecraft had been lost because the rocket scientists working on the mission saw what they wanted to see, instead of seeing what was actually in front of them.

In the last chapter, we explored how to refine and reframe the ideas you generated in the first part of the book ("Launch") by asking better questions and finding better problems. In this chapter, we'll take those refined ideas and learn how to stress-test them. I'll reveal the rocket scientist's tool kit for spotting flaws in your decision making, rooting out misinformation, and detecting errors before they snowball into catastrophe. You'll learn the test of a first-rate intelligence and the one question that will make you a better problem solver. I'll explain why a simple change in vocabulary can make your mind more flexible and what you can learn from a basic puzzle that 80 percent of people fail to solve. We'll explore the benefits of switching our default from convincing others that we're right to convincing ourselves that we're wrong.

Facts Don't Change Minds

As a former scientist, I've been trained to rely on objective facts. For years, when I was attempting to persuade someone, I would back my arguments with hard, cold, irrefutable data and expect immediate results. Drowning the other person with facts, I assumed, was the best way to prove that climate change is real, the war on drugs has failed, or the current business strategy adopted by your risk-averse boss with zero imagination isn't working.

But I've discovered a significant problem with this approach. It doesn't work.

The mind doesn't follow the facts. Facts, as John Adams put it, are stubborn things, but our minds are even more stubborn. Doubt isn't always resolved in the face of facts for even the most enlightened among us, however credible and convincing those facts might be. The same brains that empower rational thinking also skew our judgments and introduce subjective contortions.

Our tendency toward skewed judgment partly results from the confirmation bias. We undervalue evidence that contradicts our beliefs and overvalue evidence that confirms them. "It [is] a puzzling thing," Robert Pirsig writes. "The truth knocks on the door and you say, 'Go away, I'm looking for the truth,' and so it goes away."[4]

As wonderful as the internet is, it has reinforced our worst tendencies. We accept as truth the first Google hit that confirms our beliefs—even if the hit appears on page 12 of the search results. We don't seek multiple references or filter out low-quality information. We quickly jump from "This sounds right to me" to "This is true."

Confirming our theories feels good. We get a hit of dopamine every time we're proven right. In contrast, hearing opposing views is a genuinely unpleasant experience—so much so that people turn down cold, hard cash to remain in their ideological bubble. In a study of over two hundred Americans, roughly two-thirds of the participants refused the opportunity to win extra money by listening to the other side's arguments on same-sex marriage.[5] They didn't turn down the money because they already knew what the other side thought. No, the participants explained to the researchers that hearing the opposing views would be too frustrating and discomforting to them. The results were ideologically neutral: Participants on both sides of the question were equally likely to refuse cash if it required listening to the other side.

When we seclude ourselves from opposing arguments, our opinions solidify, and it becomes increasingly harder to disrupt our established patterns of thinking. Aggressively mediocre corporate managers remain employed because we interpret the evidence to confirm the accuracy of our initial hiring decision. Doctors continue to preach the ills of dietary cholesterol despite research to the contrary. University students maintain their beliefs even when those beliefs violate the laws of physics.

Recall that it was Galileo who discovered, through a thought experiment, that objects of different masses fall at the same rate in a vacuum. In one study, university students were asked whether they thought heavier objects fell faster than lighter ones.[6] After they recorded their answers, the students then observed a physical demonstration where a metal and a plastic of the same size were dropped from the same height in a vacuum. Although the two objects fell at the same rate, the students who initially believed the heavier metal would fall faster were more likely to report that the metal did fall faster.

In a different study, researchers sent more than 1,700 parents one of four campaigns intended to increase vaccination rates for measles, mumps, and rubella (MMR).[7] The campaigns, which were adopted nearly verbatim from those used by federal agencies, took different approaches. For example, one campaign offered textual information refuting the vaccine-autism link, and another showed graphic images of children who had developed diseases that could have been prevented by the vaccine. The study's goal was to determine which campaign would be the most effective in overcoming parents' reluctance to vaccinate their children.

Remarkably, none of the campaigns worked. For parents with the least favorable attitude toward vaccines, the campaigns actually backfired and made the parents *less* likely to vaccinate their children. For already-hesitant parents, the fear-based campaign—bearing tragic images of children suffering from measles—paradoxically *increased* the belief that the MMR vaccine causes autism. The graphic images might have caused nervous parents to think of additional dangers to their children—dangers that they then associated with vaccines. "The best response to false beliefs," the researchers concluded, "is not necessarily providing correct information."

You may be thinking, facts may not trump emotion in parents, but the same can't be true for rocket scientists—the brilliant class of rational people entrusted with expensive spacecraft precisely because they have been trained to make sound judgments based on objective data. As we'll see in the next section, however, even rocket scientists can have a hard time thinking like a rocket scientist.

Something Funny's Going On

With a smartphone in most pockets, navigation problems are largely a thing of the past. Gone are the days when we rolled down our car window to ask for directions to a wholesome-looking stranger and, when those directions inevitably led us astray, asked other strangers to course-correct. Now, we simply plug in our destination and instantaneously get a play-by-play of our route.

Navigating interplanetary spacecraft, however, feels more like old-fashioned driving. There's no window rolling involved, but during launch and the ensuing flight, the spacecraft picks up inaccuracies in its trajectory. These inaccuracies are expected in each flight, so the navigation team schedules trajectory correction maneuvers, firing the spacecraft's engines to ensure it remains en route—the equivalent of asking additional strangers for directions along the way.

For the Mars Climate Orbiter, four trajectory correction maneuvers were planned by the group of engineers responsible for the spacecraft's navigation at JPL.[8] During the fourth maneuver, which occurred roughly two months before the spacecraft's arrival at Mars, something strange happened. The data collected after the burn showed that the spacecraft's altitude would be lower than expected when it entered Martian orbit. The downward drift was subtle, but palpable and continuous. As the spacecraft moved closer to Mars, it inexplicably kept creeping down.

Some predictions were off by as much as seventy kilometers from the aim point. Yet the "navigators still acted as if they believed the aiming accuracy was within 10 kilometers."[9] This seventy-kilometer discrepancy, according to one expert, "should have [had] people screaming down the halls. This tells you that you have no idea where your spacecraft is, and therefore your trajectory has an unacceptable probability of intersecting the planet's atmosphere."[10] Still, the navigators assumed the error was with the navigation software, not the spacecraft's trajectory, which seemingly remained *nominal*—rocket-science lingo for "as expected."

There were murmurs within JPL that all wasn't nominal with the orbiter. A week or two before the orbiter's scheduled entry into Martian orbit, Mark Adler checked in with the members of the orbiter team to see how things were going (You might recall Adler from an earlier chapter. He's the JPL engineer who came up with the airbag idea for the Mars Exploration Rovers). Adler kept receiving the same cryptic response: "Something funny's going on." But the navigators appeared confident. "It will sort itself out," they told Adler.

Although there were only four planned trajectory correction maneuvers, there remained the possibility of adding a fifth. But the team members decided to pass. They continued to believe that the spacecraft would enter Mars at a safe altitude—despite the data screaming otherwise.

What ultimately happened to the orbiter takes me back to my high-school physics class. Our teacher would give us zero points for an exam answer that lacked units of measurement. She had no mercy: Even if the response was correct, we'd fail if we wrote down "150" instead of "150 meters." I had a laissez-faire approach to units of measurement and didn't get why they were such a big deal—until I learned more about the navigation error that killed the Mars Climate Orbiter.

It turned out that Lockheed Martin, which built the orbiter, was using the English inch-pound system, but JPL, which navigated the orbiter, was using the metric system. When Lockheed programmed a piece of trajectory software, JPL engineers assumed—incorrectly, it turned out—that the numbers were in newtons, the metric unit of force. One pound of force is 4.45 newtons, so the relevant measurements were all off by more than a factor of four. JPL and Lockheed Martin were speaking different languages, and neither team was aware of the problem, because both groups forgot to include units of measurement.

These rocket scientists would have all failed my high-school physics class.

But to chalk up this $193 million catastrophe to NASA's inability to do high-school physics or Lockheed Martin's inexplicable use of the archaic inch-pound system would be to vastly oversimplify the matter. The rocket scientists working on the project fell victim to the same biases that detract all humans from rational thinking. "People sometimes make errors," NASA associate administrator Edward Weiler explained after the orbiter's crash. "The problem here was not the error. It was the failure of NASA's systems engineering, and the checks and balances in our processes to detect the error. That's why we lost the spacecraft." There was a gap—which went undetected—between the story the data told and the story the rocket scientists told themselves.

No one comes equipped with a critical-thinking chip that diminishes the human tendency to let personal beliefs distort the facts. Regardless of your intelligence, Feynman's adage holds true: "The first principle is that you must not fool yourself—and you are the easiest person to fool."[11]

Instead of resenting their genetic wiring, scientists have come up with a set of tools to correct for their all-too-human inclination to fool themselves. These tools aren't just for the scientist. Rather, they're a set of tactics—an assembly of trajectory correction maneuvers—that we can all use to stress-test our ideas and spot the truth.

We begin in an unlikely place—a work of fiction—that provides a remarkably faithful glimpse at the scientist's critical-thinking tool kit: a scene from the movie *Contact*.

The Case Against Opinions

It's dusk in the middle of the New Mexico desert. Jodie Foster's character, Ellie Arroway, is a scientist searching for extraterrestrial life. She's lying on top of her car, with the white dish-shaped antennas of the Very Large Array spinning in the background. Arroway's eyes are closed, headphones plugged in, and the rest of the world tuned out. She's listening for radio signals from outer space—waiting for E.T. to call.

Just as she's settling in, a loud, rhythmic signal rises above the cosmic noise and jolts her awake. "Holy shit," she blurts out. She hops into her car, starts screaming coordinates and instructions at her oblivious coworkers over a walkie-talkie, and speeds wildly back to the office.

Once Arroway is back at the office, the team there jumps into action, moving equipment around, turning knobs, checking frequencies, and typing stuff into various different computers.

"Make me a liar, Fish!" Arroway yells at her colleague Fisher.

Fisher then starts spinning various alternative hypotheses about the source of the signal. "It could be AWACS out of Kirkland jamming us," he says, referring to the Airborne Warning and Control System. But the status of AWACS is negative, so that's ruled out. Other possible sources are also checked off. "NORAD's not tracking any snoops in this vector," Fisher says, referring to the North American Aerospace Defense Command, before adding that the Space Shuttle *Endeavor* is also in sleep mode. Arroway then checks FUDD—short for the Follow-Up Detection Device—used to confirm that the signal came from space and not from Earth. Once the space origin is confirmed, she kisses the computer screen and says, "Thank you, Elmer!" while delighting Looney Tunes lovers everywhere.[12]

The point source of the signal is later determined. It's the star Vega. But instead of settling on the answer, the team immediately moves to prove this hypothesis wrong: Vega is too close, it's too young to have developed intelligent life, and it has been scanned a bunch of times before with negative results.

But the signal is unmistakable. Soon enough, they realize that the signal is a sequence of prime numbers—a clear sign of intelligence. For a moment, Arroway contemplates immediately going public but then reconsiders. She knows that the discovery has to be independently confirmed and replicated by other scientists. The signal could be a spoof, a glitch, a delusion—any number of things could have led her American team astray.

So she goes global. Because Vega is quickly setting in the United States, she dials a colleague in Parks Observatory, which hosts a radio telescope in New South Wales, Australia. The Aussie colleague confirms the signal.

"Do you have a source location yet?" Arroway asks, without revealing her own findings.

"We put it right smack in the middle," the Australian replies. After a brief pause that seems to last for minutes, he adds, "Vega."

Arroway steps back from her computer, letting the magnitude of the moment set in.

"Who're we going to call now?" asks a coworker.

"Everybody," Arroway says.

....

THIS SCENE—WHICH we're going to dissect in the remainder of this chapter—is based on a novel authored by Carl Sagan, whose scientific touch is unmistakable. Yes, the movie's director, Robert Zemeckis, took some scientific liberties. Most obviously, scientists don't listen to radio signals on their headphones in the middle of the desert. They use computers. ("I had to take license here," Zemeckis explained. "It's only a romantic image."[13]) But ventures into cinematic romance are rare in this scene.

The first thing to note is what Arroway does *not* do. Even when she hears a distinct signal that appears to be a sign of intelligent life, she refrains from immediately blurting out an opinion about what the signal might mean.

From a scientific perspective, opinions present several problems. Opinions are sticky. Once we form an opinion—our own very clever idea—we tend to fall in love with it, particularly when we declare it in public through an actual or a virtual megaphone. To avoid changing our mind, we'll twist ourselves into positions that even seasoned yogis can't hold.

Over time, our beliefs begin to blend into our identity. Your belief in CrossFit makes you a CrossFitter, your belief in climate change makes you an environmentalist, and your belief in primal eating makes you paleo. When your beliefs and your identity are one and the same, changing your mind means changing your identity—which is why disagreements often turn into existential death matches.

As a result, at the outset of their investigation, scientists refrain from stating opinions. Instead, they form what's called a working hypothesis. The operative word is *working*. Working means it's a work in progress. Working means it's less than final. Working means the hypothesis can be changed or abandoned, depending on the facts.

Opinions are defended, but working hypotheses are tested. The test is performed, as geologist and educator T. C. Chamberlin explains, "not for the sake of the hypothesis, but for the sake of facts."[14] Some hypotheses mature into theories, but many others don't.

In my early years in academia, I ignored all the advice I'm shelling out here. I treated my papers as final opinions, rather than working hypotheses. Whenever someone challenged one of my opinions during an academic presentation, I'd get defensive. My heart rate would skyrocket, I would tense up, and my answer would reflect the annoyance with which I viewed the question and the questioner.

I then went back to my scientific training and began to reframe my opinions as working hypotheses. I changed my vocabulary to reflect this mental shift. At conferences, instead of saying "I argue . . . ," I began to say "This paper hypothesizes. . . ."

In my case, this subtle verbal tweak tricked my mind into separating my arguments from my personal identity. Obviously, I was the one who came up with the ideas, but once they were out of my body, they took on a life of their own. They became separate, abstract things I could view with some objectivity. It was no longer personal. It was a working hypothesis that simply needed more work.

But even a working hypothesis is an intellectual child that can generate emotional attachment. One remedy, as we'll see in the next section, is to have multiple children.

A Family of Hypotheses

Radio telescopes are used not only to scope out alien life as in *Contact* but also to make long-distance, interplanetary phone calls to spacecraft traveling through the solar system.[15] The Deep Space Network—a combination of three giant radio antenna arrays—serves as the hub of this network. The tracking stations are spread equidistantly from each other across the globe in Goldstone, California; near Madrid, Spain; and near Canberra, Australia. As the Earth rotates and one station loses a signal, the next picks up the baton.

On December 3, 1999, the Madrid station was tracking the Mars Polar Lander as it barreled toward the Martian surface on the night of its scheduled landing. The lander was arriving at Mars a few months after the embarrassing loss of the Mars Climate Orbiter from the mismatch in units of measurement. This was NASA's opportunity to save face.

At about 11:55 a.m. Pacific Time, the lander entered the Martian atmosphere and began its descent to the surface. As scheduled, the Madrid station lost the signal from the lander. If everything went according to plan, the Goldstone station would pick up the signal again at 12:39 p.m.

But 12:39 p.m. came and went with no word from the lander. The search for a signal continued for several days, with engineers repeatedly beaming commands to the lander. Their calls went unreturned.

Just when NASA was about to pronounce the lander dead, something strange happened. On January 4, 2000, after a month of silence from the lander, a signal from Mars was picked up by an extremely sensitive radio telescope at Stanford University. "It was the radio-frequency equivalent of a whistle," explained Ivan Linscott, a senior research associate at

Stanford.[16] The whistle had the exact characteristics you would expect of a signal from the Mars Polar Lander. To verify the signal's origins, the scientists told the spacecraft to send smoke signals by turning its "radio on and off in a distinctive sequence."[17] The spacecraft appeared to oblige. The scientists received the smoke signal and announced, much like Dr. Frankenstein, that the spacecraft was alive.

Except it wasn't. The signal turned out to be a fluke. The Stanford scientists were experiencing a phenomenon known as "I wouldn't have seen it if I hadn't believed it."[18] Radio telescopes in the Netherlands and the United Kingdom attempted to locate the signal but couldn't replicate the Stanford results.

The problem was diagnosed by Francis Bacon nearly four centuries ago: "It is the peculiar and perpetual error of the human understanding to be more moved and excited by affirmatives than negatives."[19] Stanford's search technique was designed to ferret out signals from the Mars Polar Lander. It's a signal the team members were expecting—no, hoping—to see. And that's exactly what they saw.

What's more, the scientists were emotionally attached to the lander's survival. "It's like having a loved one missing in action," explained JPL research scientist John Callas.[20] Desperately wanting to believe the lander was alive, they concluded that it was.

This wasn't the first time scientists were hoodwinked by imaginary signals from Mars. Tesla also reported detecting messages that came from Mars and consisted of a "regular repetition of numbers," similar to Arroway's detection of prime numbers from Vega. Tesla interpreted these numbers as "extraordinary experimental evidence" of intelligent life on Mars.[21]

None of these scientists were intentionally trying to mislead the public. Their conclusions were based on their interpretation of seemingly objective data. So how did these brilliant people see something when there was nothing?

A hypothesis—even a working one—is still an intellectual child. As Chamberlin explains, the hypothesis "grows more and more dear to [its author], so that, while he holds it seemingly tentative, it is still lovingly tentative, and not impartially tentative. . . . From an unduly favored child, it readily becomes master, and leads its author whithersoever it will."[22]

When we start with a single hypothesis and run with the first idea that pops into mind, it's much easier for that hypothesis to become our master. It anchors us and blinds us to alternatives sitting in the periphery. As author Robertson Davies put it, "The eye sees only what the mind is prepared to comprehend."[23] If the mind anticipates a single answer—the Mars Polar Lander may be alive—that's what the eye will see.

Before announcing a working hypothesis, ask yourself, what are my preconceptions? What do I believe to be true? Also ask, do I really want this particular hypothesis to be true? If so, be careful. Be very careful. Much as in life, if you like someone, you'll tend to overlook their flaws. You'll find signals from a love interest—or a spacecraft—even when they're not sending any.

To make sure you don't fall in love with a single hypothesis, generate several. When you've got multiple hypotheses, you reduce your attachment to any one of them and make it more difficult to quickly settle on one. With this strategy, as Chamberlin explains, the scientist becomes "the parent of a family of hypotheses: and, by his parental relation to all, he is forbidden to fasten his affections unduly upon any one."[24]

Ideally, the hypotheses you spin should conflict with each other. "The test of a first-rate intelligence," F. Scott Fitzgerald said, "is the ability to hold two opposed ideas in mind at the same time, and still retain the ability to function."[25] This approach isn't easy. Even scientists can have a hard time entertaining multiple viewpoints without having their heads explode. For centuries, the scientific community was split into two camps, with one camp believing light is a particle like motes of dust and others arguing it's a wave, like the ripples in the water. It turned out that both camps were right (or wrong, depending on how you view it). Light straddled these two categories and exhibited the properties of both a particle and a wave.

The Large Hadron Collider is a seventeen-mile particle accelerator that smashes together subatomic particles called hadrons. Their collision has been described as "less of a collision and more of a symphony."[26] When hadrons collide, they actually glide through each other, and "their fundamental components pass so close together that they can talk to each other."[27] If this symphony plays out the right way, the colliding hadrons "can pluck deep hidden fields that will sing their own tune in response—by producing new particles."[28]

Multiple hypotheses dance with each other the same way. If you can hold conflicting thoughts in your head and let them dance with each other, they'll produce a symphony that will bring out additional notes—in the form of new ideas—far superior to the original ones.

But how do you generate conflicting ideas? How do you find the countermelody to your melody? One approach is to actively look for what's missing.

What's Missing?

The twenty-seven-year-old director had a major problem in his hands.[29] Bruce, the star of his movie, was high maintenance, even by Hollywood standards. Bruce was a mechanical shark, lovingly named after the director's lawyer. But the shark couldn't do the one thing it was built to do: swim properly. On its first day on the set, the shark sank to the bottom of the water. Within a week, its electric motor malfunctioned. Even after a good day, Bruce "had to be drained, scrubbed, and repainted" to get ready for filming—requiring the type of pampering rarely expected by movie stars.

The director then did what all directors wish they could do to an overly demanding and underperforming actor. He fired the shark. "I had no choice but to figure out how to tell the story without the shark," he explained. As he faced this major constraint, he asked himself, "What would Hitchcock do in a situation like this?" The answer gave him a stroke of inspiration that helped him convert a seemingly insurmountable obstacle into a blockbuster opportunity.

In the opening scene of the movie, Chrissie decides to go for a moonlight dip. As she's swimming along, she is suddenly pulled underwater and yanked around, while gasping for breath and screaming for help. The focus is on Chrissie, and the villain is nowhere to be seen. The monster is left entirely to the imagination of the audience, which doesn't get a good look at the shark until the third act. This omission ultimately produced a constant state of anxiety in the audience—a feeling boosted by the ominous theme music (*da-dum . . . da-dum . . . da-dum-da-dum-da-dum*).

The movie, as you probably guessed, was *Jaws*, and its director was a young Steven Spielberg. Even early in his career, Spielberg knew what many of us neglect to acknowledge: What we don't see can be scarier than what we do see.

From a human perspective, not all facts are equal. We tend to incessantly focus on the facts in front of us and neglect other facts that may be hidden in a blind spot.

This blind spot results, in part, from our genetic programming. As psychologist Robert Cialdini explains, "It is easier to register the presence of something than its absence."[30] We're wired to respond to the obvious signs: the rattling in the dark, the smell of gas, the sight of smoke, the screeching of tires. Our pupils dilate, our heart starts pumping faster, and adrenaline is released. Our minds zero in on the potential threat, filtering all other sensory inputs. These mechanisms are essential to our survival, but they also supersede other operations and cause us to miss crucial pieces of data.

In one famous study, researchers filmed a group of six people—half wearing white and the other half wearing black shirts—passing a basketball to each other. The instructions were simple (dare I say, not rocket science): "Count how many times the players wearing white pass the basketball." Roughly ten seconds into the film, a person wearing a gorilla costume slowly walks into the frame. She conspicuously stops in the middle of the players, faces the camera, beats her chest while the players continue to pass the ball around her, and then exits out of the frame. This isn't a subtle interruption—the gorilla appears impossible to miss.[31] Yet, half the participants in the study didn't see it at all. They were so preoccupied with counting the passes that they ignored the gorilla in the room.[32]

But contrary to popular wisdom, what you don't see or know can hurt you. The amateur lawyer doesn't see the winning legal argument. The mediocre doctor misses the right diagnosis. The average driver doesn't realize where the potential dangers lie.

In focusing on the facts in front of us, we don't focus enough—or at all—on the missing facts. As the focal facts scream for attention, we must ask, "What am I not seeing? What fact should be present, but is not?" Follow the lead of the scientists from *Contact*, where they repeatedly asked themselves what they could be missing—the signal could have come from AWACS, NORAD, or the Space Shuttle *Endeavor*.

The rocket scientists working on the Mars Climate Orbiter neglected to ask these questions. An invisible force kept tugging at the orbiter—just as it tugged at Chrissie—and kept bringing it downward as it swam across the cosmic ocean. But the shark that was the mismatch in units of measurement remained hidden. Despite the warning signs, no one formally raised their hand and asked, "Are we missing something?"

A postmortem conducted after the orbiter's crash advocated that team members adopt "a Sherlock Holmes approach and a bulldog's disposition to pursue strange indications."[33] The team had built a theory without gathering all the facts—which, if you know your Holmes, is the worst mistake an investigator can make—and then refused to let the facts disturb it.

The importance of searching for the hidden facts is central to the mystery story "Silver Blaze," where Holmes reveals a theft to be an inside job by focusing on what's missing:

GREGORY (Scotland Yard detective): Is there any other point to which you would wish to draw my attention?
HOLMES: To the curious incident of the dog in the nighttime.
GREGORY: The dog did nothing in the nighttime.
HOLMES: That was the curious incident.[34]

The dog guarding the property hadn't barked, so Holmes concluded that the thief couldn't have been the stranger the police had rushed to lock up.

So, the next time you're tempted, my dear Watson, to announce a confident conclusion, do what you do every time you drive. Don't simply rely on the visible hazards from the rearview and side mirrors. Ask yourself, "What's missing?" When you think you've exhausted all possibilities, keep asking, "What else?" Make a deliberate effort to repeatedly turn your head and check your blind spot.

You'll be surprised to find sharks lurking there.

....

FINDING WHAT'S MISSING and using that information to generate multiple hypotheses is helpful, but it doesn't guarantee objectivity. Unwittingly, you may give the benefit of the doubt to one of your intellectual offspring for a postcurfew arrival but ground the others for the very same offense. This is why, after you've spun multiple intellectual darlings, you must do the unthinkable: kill them off.

Kill Your Intellectual Darlings

An experimenter walks into a room and gives you these three numbers: 2, 4, 6. She tells you that the numbers follow a simple rule, and your job is to discover the rule by proposing different strings of three numbers. The experimenter will then tell you whether the strings you propose conform to the rule. You get as many tries as you want, and there's no time limit.

Give it a shot. What do you think the rule is?

For most participants, the experiment went in one of two ways. Participant A said "4, 6, 8." The experimenter replied, "Follows the rule." The participant then said, "6, 8, 10." The experimenter said, "Also follows the rule." After several more strings of numbers were met with nods of approval, Participant A declared that the rule was "increasing intervals of two."

Participant B opened with "3, 6, 9." The experimenter replied, "Follows the rule." The participant then said, "4, 8, 12." The experimenter's response: "Also follows the rule." After Participant B produced several more strings of numbers that conformed to the rule, he declared that the rule was "multiples of the first number."

Much to their astonishment, both participants were wrong.

The rule, it turns out, was "numbers in increasing order." The strings of numbers that both Participant A and Participant B provided conformed to the rule, but the rule was different from the one they had in mind.

If you didn't get the rule right, you're in good company. Only about one in five people in the study could identify the rule on their first attempt.

What's the secret to solving the puzzle? What set apart the successful participants from the unsuccessful ones?

The unsuccessful participants believed they found the rule early on and proposed strings of numbers that confirmed their belief. If they thought the rule was "increasing intervals of two," they generated strings like 8, 10, 12 or 20, 22, 24. As the experimenter validated each new string, the participants grew increasingly more confident in their initial brilliant hunch and assumed they were on the right track. They were too busy trying to find numbers that conformed to what they thought was the right rule, rather than discovering the rule itself.

The successful participants took the exact opposite tack. Instead of trying to prove themselves right by generating strings that confirmed their hypothesis, they tried to falsify it. For example, if they thought the rule was "increasing intervals of two," they would say, "3, 2, 1." That string doesn't follow the rule. They might then say, "2, 4, 10." That string follows the experimenter's rule, but doesn't follow what most participants assumed was the right rule.

The numbers game, as you may have guessed, is a microcosm for life. Our instinct in our personal and professional lives is to prove ourselves right. Every yes makes us feel good. Every yes makes us stick to what we think we know. Every yes gets us a gold star and a hit of dopamine.

But every no brings us one step closer to the truth. Every no provides far more information than a yes does. Progress occurs only when we generate negative outcomes by trying to rebut rather than confirm our initial hunch.

The point of proving yourself wrong isn't to feel good. The point is to make sure your spacecraft doesn't crash, your business doesn't fall apart, or your health doesn't break down. Each time we validate what we think we know, we narrow our vision and ignore alternative possibilities—in the same way each nod of approval from the experimenter led participants to fixate on the wrong hypothesis.

The numbers study is from a real experiment conducted by cognitive psychologist Peter Cathcart Wason, who coined the term *confirmation bias*.[35] Wason was interested in exploring what Karl Popper had termed *falsifiability*, which means that scientific hypotheses must be capable of being proven wrong.[36]

Take, for example, the statement "All doves are white." This statement is falsifiable. If you find a black dove, or a brown dove, or a yellow dove, you've proven the hypothesis wrong—similar to how a nonconforming string of numbers can falsify your initial hunch in the numbers study.

A scientific theory is never proven right. It's simply not proven wrong. Only when scientists work hard—but fail—to beat the crap out of their own ideas can they begin to develop confidence in those ideas. Even after a theory gains acceptance, new facts often emerge, requiring the refinement or complete abandonment of the status quo.

"Nothing in the physical world seems to be constant or permanent," physicist Alan Lightman writes. "Stars burn out. Atoms disintegrate. Species evolve. Motion is relative."[37] The same is true for facts. Most facts have a half-life. What we're advised with confidence this year is reversed the next.

The history of science, as clinician and author Chris Kresser says, "is the history of most scientists being wrong about most things most of the time."[38] Aristotle's ideas were falsified by Galileo's, whose ideas were replaced by Newton's, whose ideas were modified by Einstein. And Einstein's own theory of relativity broke down at the subatomic level—in the imperceptible land of tiny particles like quarks, gluons, and hadrons—where quantum field theory now rules. We were certain about each of these facts—until we were not. The "here today, gone tomorrow" nature of scientific theory is simply its "natural rhythm," Gary Taubes writes.[39]

Although scientists dedicate their lives to cross-examining their own ideas, this mode of operation runs counter to human conditioning. In politics, for example, consistency trumps accuracy. When politicians admit to changing their minds—because the facts have changed or a better argument persuaded them—they are castigated by the opposition for flip-flopping. They are dragged through the mud for being inconsistent, indecisive, and generally unfit to be the hard, ideological person suitable for elected office.

To most politicians, the statement "this argument is irrefutable" is a virtue. But to scientists, it's a vice. If there's no way to test a scientific hypothesis and disprove it, it's essentially worthless. As Sagan explains, "Skeptics must be given the chance to follow your reasoning, to duplicate your experiments, and see if they get the same result."[40]

Consider, for example, the "simulation hypothesis," first posited by philosopher Nick Bostrom and later popularized by Elon Musk. The hypothesis says we're little creatures living in a computer simulation controlled by more-intelligent powers.[41] This hypothesis isn't falsifiable. If we're like the characters in the video game *The Sims*, we can't acquire information about our world from outside it. As a result, we can never prove that our world is *not* just an illusion.

Falsification is what separates science from pseudoscience. When we keep opposing arguments at bay through unfalsifiable arguments and disable others from challenging our beliefs, misinformation thrives.

Once we create falsifiable hypotheses, we must follow the successful participants in the numbers study and attempt to falsify these hypotheses, rather than searching out information to prove them right. Ideological lock-in happens without our awareness. We must therefore deliberately expose ourselves to the discomfort of self-falsification instead of merely repeating platitudes like "I'm open to proving myself wrong." When our focus shifts from proving ourselves right to proving ourselves wrong, we seek different inputs, we combat deeply entrenched biases, and we open ourselves up to competing facts and arguments. "I don't like that man,"

Abraham Lincoln is said to have observed. "I must get to know him better." The same approach should apply to opposing arguments.

Regularly ask yourself—as Stewart Brand, the founder of the *Whole Earth Catalog* does—How many things am I dead wrong about?[42] Poke holes in your most cherished arguments, and look for disconfirming facts (What fact would change my mind?). Follow the "golden rule" of Darwin who, upon finding a fact that contradicted one of his beliefs, would write it down right away.[43] When you kill off your bad or outdated ideas, Darwin knew, you leave breathing room for the good ones to surface. By making you question deeply entrenched beliefs, this tactic can also boost first-principles thinking.

Consider also Daniel Kahneman, who won the Nobel prize in 2002 for his groundbreaking work on the psychology of judgment and decision making. Taking home the Nobel is an impressive feat, but it's all the more impressive in Kahneman's case. He won the prize for economics, and he's a psychologist. "Most people after they win the Nobel Prize just want to go play golf," explained Princeton professor Eldar Shafir. "Danny's busy trying to disprove his own theories that led to the prize. It's beautiful, really."[44] Kahneman also invites his critics to join in on the fun by persuading them to collaborate with him.[45]

One of my favorite US Supreme Court opinions is Justice John Marshall Harlan's dissenting opinion in the 1896 case of *Plessy v. Ferguson*. In that case, a majority of the court, against Harlan's sole dissent, upheld the constitutionality of racial segregation (the case was later overturned in *Brown v. Board of Education*).

Harlan's dissent came as a surprise to many. Harlan was a white supremacist. He used to own slaves.[46] He staunchly opposed the Reconstruction Amendments to the US Constitution, which prohibited the government from discriminating on the basis of race (among other things). When Harlan's critics accused him of flip-flopping, his answer was simple:

I'd rather be right than consistent.[47]

"One mark of a great mind," Walter Isaacson said, "is the willingness to change it."[48] When the world around you changes—when the tech bubble bursts or self-driving cars become the norm—the ability to change with the world confers an extraordinary advantage. "The successful executive is faster to recognize the bad decisions and adjust," explains Walt Bettinger, the CEO of Charles Schwab, "whereas failing executives often dig in and try to convince people that they were right."[49]

If you have trouble challenging your own beliefs, you can pretend they're someone else's. In writing this book, I adopted a strategy from Stephen King, who puts away his draft chapters for weeks before returning to them. When he comes back with some psychological separation, it's easier to pretend that someone else wrote the chapter. Seeing the work from a fresh perspective removes his blinders and enables him to hack away at the writing. King's approach finds support from research. In one study, participants became more critical of their own ideas when those ideas were presented to them as if they were someone else's.[50]

In the end, if we don't prove ourselves wrong, others will do it for us. If we pretend to have all the answers, our cover will eventually be blown. If we don't recognize the flaws in our own thinking, those flaws will come to haunt us. As cognitive scientists Hugo Mercier and Dan Sperber point out, a mouse "bent on confirming its belief that there are no cats around" will end up as food for those cats.[51]

Our goal should be to *find* what's right—not to *be* right.

Years after he published the numbers study that opened this section, Wason was stopped on the street by Imre Lakatos, a philosopher of science at the London School of Economics. "We've read everything you have written," Lakatos told Wason, "and we disagree with all of it." He added, "Do come and give us a seminar."[52]

In extending an invitation to his intellectual opponent, Lakatos was following a strategy we'll explore in the next section.

A Light-Filled Box

Niels Bohr and Albert Einstein were among the greatest intellectual rivals in science. They engaged in a series of public debates about quantum mechanics—specifically about the uncertainty principle, which says that it's impossible to determine both the exact position and the exact momentum of subatomic particles.[53] Bohr supported this principle, but Einstein opposed it.

Despite their sharp intellectual disagreements, the relationship between Bohr and Einstein was one of mutual respect. True to form, Einstein came up with a series of thought experiments to challenge the uncertainty principle. During the Solvay conferences on physics, which brought together the world's most prominent physicists, Einstein would arrive for breakfast and giddily announce he had invented another thought experiment that falsified the uncertainty principle.[54]

Bohr would consider Einstein's challenge all day. By dinner, Bohr would usually have an answer to put Einstein in his place. Einstein would then retreat to his hotel room and descend to breakfast the next day armed with a brand-new thought experiment.

This intellectual boxing was like Rocky Balboa and Apollo Creed sparring after hours at the gym—two giants, tuning the world out, testing their craft on each other, and growing stronger as a result. In each man's work, you can see traces of the other—if not in name, certainly in spirit. It's not about winning or losing. It's about the game or, in this case, the science.

Bohr and Einstein turned to each other to stress-test their opinions because the men were too close to their perspectives to see their own blind spots. "One thing a person cannot do, no matter how rigorous his analysis or heroic his imagination," Nobel laureate Thomas Schelling once observed, "is to draw up a list of things that would never occur to him." This is why in *Contact*, Arroway yells "Make me a liar, Fish," asking her colleague to prove her wrong.[55]

This is also why disagreement is built into the scientific process. "Progress in science," theoretical physicist John Archibald Wheeler says, "owes more to the clash of ideas than the steady accumulation of facts."[56] Even scientists who work in seclusion must eventually expose their ideas to their colleagues through peer review—a hurdle that all major scientific publications must clear. But publication isn't the end of it. The conclusions in the publication must then be independently verified by other scientists who have no motive to support the ideas—in the same way that the sequence of prime numbers in *Contact* was verified by Arroway's Australian colleagues.

In one of my favorite commencement speeches of all time, David Foster Wallace tells the story of two young fish. The fish are swimming along, "and they happen to meet an older fish swimming the other way, who nods at them and says, 'Morning, boys, how's the water?'" The two young fish swim on, "and then eventually one of them looks over at the other and goes, 'What the hell is water?'"[57]

Everything we observe in the world is through our own eyes. What may be obvious to others—we're swimming in water—isn't obvious to us. Others have that seemingly freakish ability to spot the mismatch in our units of measurement or our collective delusion about a signal from a dead Martian lander. They're not wedded to our vision of the world, they don't have the same emotional attachment to our opinions, and they won't swat away conflicting information like we do. "The road to self-insight," psychologist David Dunning said, "runs through other people."[58]

Yet this road is often obstructed. In the modern world, we live in a perpetual echo chamber. Although technology has torn down some barriers, it has ended up erecting others. We friend people like us on Facebook. We follow people like us on Twitter. We read blogs and newspapers that vibrate on the same political frequency. It's easy to connect only with our tribe and disconnect from the others. Just unsubscribe, unfollow, or unfriend.

This internet-fueled tribalism exacerbates our confirmation bias. As our echo chambers get louder and louder, we're repeatedly bombarded with ideas that reiterate our own. When we see our own ideas mirrored in others, our confidence levels skyrocket. Opposing ideas are nowhere to be seen, so we assume they don't exist or that those who adopt them must be irrational.

As a result, we must consciously step outside our echo chamber. Before making an important decision, ask yourself, "Who will disagree with me?" If you don't know any people who disagree with you, make a point to find them. Expose yourself to environments where your opinions can be challenged, as uncomfortable and awkward as that might be. If you're Niels Bohr, who is your Albert Einstein lobbing thought experiments at you? If you're Ruth Bader Ginsburg, where's your Antonin Scalia writing a cheeky but powerful dissent? If you're Andre Agassi, who is your Pete Sampras to keep you on your toes with a powerful serve?

You can also ask people who normally agree with you to disagree with you. For example, I gave trusted advisers early drafts of this book and asked them to point out not what's right, not what they loved, but what's wrong, what should be changed, what should be taken out. This approach provides psychological safety to those who might otherwise withhold dissent for fear of offending you.

If you can't find opposing voices, manufacture them. Build a mental model of your favorite adversary, and have imaginary conversations with them. This is what Marc Andreessen does. "I have a little mental model of Peter Thiel," explains Andreessen, referring to fellow venture capitalist and PayPal cofounder, "a simulation that lives on my shoulder, and I argue with him all day long."[59] He added, "People might look at you funny while it's happening," but it's well worth the ridicule.

The voice of dissent could be anyone. You can ask yourself, "What would a rocket scientist do?" and imagine a rocket scientist, armed with the tools in this book, critically questioning your ideas. Think through what a dissatisfied customer would say about your newest product or how a new CEO who would replace you would approach the same problem (a trick that former Intel CEO Andy Grove used).[60]

In constructing a model of how an adversary thinks, you must be as objective and fair as possible. Avoid the instinct to caricature the opposing position, making it easier to debunk—a tactic called the straw man. For example, a political candidate advocates increased regulation on greenhouses gases from cars. Another candidate responds that cars are essential for people to get to work and that the proposal will destroy the economy. The argument is a straw man because the proposal calls for increased regulation—not elimination of cars—but it's far simpler to rebut the more extreme version of the idea.

Instead of using a straw man tactic, engage in its opposite, the steel man. This approach requires you to find and articulate the strongest, not the weakest, form of the opposition's argument. Charlie Munger, vice chairman of Berkshire Hathaway, is a major proponent of this idea. "You're not entitled to take a view," he cautions, "unless and until you can argue better against that view than the smartest guy who holds that opposite view."[61]

The intellectual chess game between Bohr and Einstein was so fruitful partly because they were masters at the steel man technique. Their game continued until Einstein's death. A few years later, when Bohr himself died, he left behind a drawing on his blackboard.[62] The drawing wasn't a grand revelation or defense of his own ideas. Rather, it was a light-filled box—part of a thought experiment that Einstein had posed to challenge Bohr.

Until his final breath on earth, Bohr embraced Einstein's challenges, believing them to make his ideas stronger, not weaker. His defense of quantum mechanics was based not on fortitude but on self-doubt.

In your own life, you should find that light-filled box—the challenge to your central belief systems—and never let it go. In the end, it takes courage, humility, and determination to find the truth instead of the convenient. But it's well worth the effort.

....

THERE'S A DIFFERENCE, as Morpheus said, between knowing the path and walking the path. Once you've stress-tested your ideas by trying to prove yourself wrong, it's now time to collide those ideas with reality in tests and experiments. As we'll see in the next chapter, however, rocket scientists take a radically different approach to both.

Visit **ozanvarol.com/rocket** to find worksheets, challenges, and exercises to help you implement the strategies discussed in this chapter.

7

TEST AS YOU FLY,
FLY AS YOU TEST

How to Nail Your Next
Product Launch or Job Interview

We don't rise to the level of our expectations.
We fall to the level of our training.

—UNKNOWN

MILLIONS OF AMERICANS had been waiting for this moment.[1] A promise made by a young president, a revolution of cosmic proportions, was about to be realized.

The launch was woefully behind schedule. Months before the official launch date, concerns had been raised about readiness. Yet, officials turned a blind eye and hoped the glaring problems would somehow correct themselves. They were advised to delay or abort the launch altogether, but they passed. Stress tests performed just one day before the launch date revealed a lingering flaw that could compromise the entire mission.

But the test results were ignored. In a rush to launch by the tight deadline, officials pulled the trigger. As the data began to flow in, the engineers' screens told a rapidly unfolding life-and-death story. They watched, with jaws dropped, as everything turned red.

Catastrophe ensued. Shortly after launch, it crashed and burned.

....

THIS WAS NO rocket launch. Instead, it was the unveiling of healthcare .gov—a centerpiece of the Affordable Care Act—a landmark piece of legislation enacted during President Barack Obama's term to provide affordable health insurance to Americans. The legislation was the promise, and the website was the fulfillment—or was supposed to be the fulfillment. Americans would use the website to shop for and purchase insurance.

Plagued by technical problems, the website crashed as soon as it launched. Users were unable to perform basic functions like create new accounts. The website miscalculated health-insurance subsidies and sent users into inescapable loops. Only six people were able to sign up for insurance on the first day of the website's operation.

How was healthcare.gov—so critical to the success of the Affordable Care Act—so badly botched? Why did a platform that cost nearly $2 billion fail to perform basic commands?

Rockets and websites are different beasts, but they have at least one thing in common. They'll crash unless you follow a cardinal rocket-science principle called *test as you fly, fly as you test*.

This chapter is about the test-as-you-fly principle. I'll explain how you can use this principle to test the ideas you generated in the first part of this book ("Launch") and to ensure they have the best shot at landing. You'll discover why we fool ourselves when we conduct tests and dress rehearsals and what to do about it. I'll reveal what you can learn from a flaw that corrupted the $1.5 billion Hubble Space Telescope and why one of the most popular consumer products of all time almost wasn't produced. You'll learn why a top comedian makes regular surprise visits to tiny comedy clubs and how a famous lawyer and a world-class obstacle-course racer use the same strategy from rocket science to excel in their fields.

The Problem with Tests

Most of our decisions in life are based not on tests, but on hunches and limited information. We launch a new product, we change careers, or we try a new marketing approach—all without a single experiment. We blame a lack of resources for skipping the testing but don't recognize the costs of new approaches that end up failing.

Even when we conduct tests, we perform superficial dress rehearsals that double as exercises in self-deception. We conduct tests—not to prove ourselves wrong, but to confirm what we believe is true. We tweak the testing conditions or interpret ambiguous outcomes to confirm our preconceptions.

Professors at Wharton and Harvard surveyed thirty-two cutting-edge retail companies to study their testing practices.[2] The researchers found that 78 percent of the firms tested new products in stores before launching them. Although that number is impressive, the testing conditions were not. According to the researchers, the companies believed "their products will sell well despite unfavorable test results" and blamed "the weather (bad or good), the poor choice of test sites, the inferior execution of tests, and other factors for suboptimal sales."[3] In other words, the retailers made the test fit their expectations rather than adjust their expectations to fit the test results.

In a well-designed test, outcomes can't be predetermined. You must be willing to fail. The test must run forward to shed light on uncertainty, rather than run backward to confirm preconceptions. Feynman said it best: "If it disagrees with experiment, it is wrong. In that simple statement is the key to science. It doesn't make any difference how beautiful your guess is, it doesn't make any difference how smart you are, who made the guess, or what his name is—if it disagrees with experiment, it's wrong."[4]

Self-deception is only part of the problem. The other part is the disconnect between testing conditions and reality. Focus groups and test audiences are often placed in artificial conditions and asked questions they would never get in real life. As a result, these "experiments" spit out perfectly polished, and perfectly incorrect, conclusions.

Rocket science offers a way forward with a deceptively simple principle: test as you fly, fly as you test. According to the principle, experiments on Earth must mimic, to the greatest extent possible, the same conditions in flight. Rocket scientists test the spacecraft as the spacecraft will fly. If the test is successful, the flight must take place under similar conditions. Any significant deviance between the test and the flight can cause catastrophe—whether it's a rocket, a government website, your job interview, or your next product.

In a proper test, the goal isn't to discover everything that can go right. Rather, the goal is to discover everything that can go wrong and to find the breaking point.

Breaking Point

The best way to determine an object's breaking point is to break it. Rocket scientists try to break the spacecraft on Earth—to reveal all its flaws—before the faults reveal themselves in space. This objective requires exposing every component, down to the screws, to the same type of shocks, vibrations, and extreme temperatures awaiting them in space. Scientists and engineers must think through all the ways that they can trick these components and lines of computer code into committing fatal errors.

This approach also has the benefit of reducing uncertainty, to return to an earlier chapter. Testing can help turn unknowns into knowns. Each test, if performed under similar conditions to flight, can teach rocket scientists something new about the spacecraft and prompt them to tweak a piece of software or hardware.

But even in rocket science, the testing conditions often aren't identical to the actual launch. There are certain things that we physically can't test for on Earth. For example, we can't mimic the exact same gravitational forces a rocket will experience during launch. We can't completely simulate what it will be like to drive on Mars. But we can come close.

When I was working on the 2003 Mars Exploration Rovers mission, we periodically took a test rover for a spin around the Mars Yard—a tennis-court-sized area at JPL filled with the same types of rocks you might find on the red planet. The test rover was lovingly named FIDO, short for Field Integrated Design and Operations.[5] We also took FIDO to places such as the Black Rock Summit in Nevada and Gray Mountain in Arizona. We put the rover through its paces to make sure it could do what it was supposed to do: avoid hazards, drill into rocks, take photos, and the like.

It's one thing to drive a Mars rover on Earth. But it's something else to operate it on Mars, where everything, from atmospheric density to surface gravity, is different from Earth. The closest you can come to Mars on Earth is Sandusky, Ohio. The small city boasts NASA's Space Power Facility, the world's largest vacuum chamber. It can simulate the conditions of space travel, including high vacuums, low pressures, and extreme temperature variations.[6]

The chamber provided the ideal environment for testing the airbags we would use to land our rovers on the Martian surface.[7] The Entry, Descent, Landing (EDL) team headed over to Sandusky to conduct some tests. They put a fake lander inside a set of airbags, pumped the vacuum chamber down to Martian pressures and temperatures, put some make-believe Martian rocks at the bottom of the chamber, and let it rip.

And rip they did. The rocks completely ripped through the bags, instantly deflating them. The resulting holes were big enough for a person to walk through. This one test revealed that the airbags we were planning to use were too weak.

One rock, ominously dubbed Black Rock, proved to be the perfect enemy. Adam Steltzner, who worked on the EDL team, described the rock as "the shape of a cow's liver, with a light ridge running along its top." It didn't look particularly dangerous on the surface, but "it reached in and caused the bladder inside the bags to rupture." Instead of dismissing Black Rock as an outlier—the type of rock we would be unlikely to bump up against on Mars—the EDL team members did the opposite.

They isolated the problem and then exaggerated it. They made replicas of Black Rock, scattered them across the chamber, and started flinging the airbags against them. Although the same airbags had successfully landed the *Pathfinder* on the Martian surface back in 1997, the success didn't mean the airbag design was flawless. Luck may have intervened and prevented what could have been a catastrophic collision with the wrong type of rock. But our mission's EDL team couldn't count on luck and had to assume the worst: a field of Black Rocks on Mars waiting to tear our airbags to pieces.

The fix came from a seemingly unlikely place: bicycles. Most bicycle tires have two layers—the outer layer and an inner tube. Even if the outer layer gets punctured by road debris, the inner tube remains intact. The EDL team compared apples and oranges, copying this design for our airbags and designing a double bladder for double protection. Even if the outer bladder failed, the airbag (and therefore the lander) would survive. The new design was tested and retested until the airbags survived the punishment.

You don't need a fancy vacuum chamber or a large budget to find the breaking point of your own widgets. You can run tests on prototypes or preliminary versions of your products or services using a representative group of customers. All it takes is a willingness to design tests for the worst-case—rather than the best-case—scenario.

Testing doesn't end after the spacecraft is launched. Even after take-off, we have to make sure that the instruments are operating properly in the unknown and volatile environment of space before we can begin trusting them.

We achieve that accuracy through a process called calibration. For example, each instrument on our Mars rovers had a calibration target. The fanciest target was built for our onboard camera, Pancam.[8] The target was a sundial mounted on the rover deck. On its four corners, the dial had four different color blocks containing different minerals, along with gray areas of varying reflectivity. Peppered across the target was the word *Mars* written in seventeen languages (just in case the little green folks don't speak English).[9] The dial depicted the orbits of Earth and Mars and bore the inscription "Two Worlds—One Sun." The center post of the sundial cast a shadow on the calibration target. Scientists used this shadow to adjust the brightness of the images.

Before we used any of the instruments, we would first point the instrument toward its calibration target. Pancam, for example, would snap a photo of the sundial and send it down to Earth. If the readings on Mars didn't match our readings of the same target on Earth—if, for example, the green block on the sundial appeared red in the calibration photo—we would know the instrument was miscalibrated.

In our daily lives, we are miscalibrated far more often than we assume. We need a calibration target, preferably multiple trusted advisers, who can warn us when our reading of the events is off—when we're looking at a green block but seeing red. Pick your calibration targets carefully, and make sure you can trust their judgment. If their judgment is off, yours will be too.

As we'll see in the next section, it's not enough to test the reliability of individual components. Without systems-level testing, you can unwittingly unleash Frankenstein's monster.

Frankenstein's Monster

In one sense, a spacecraft is no different from your business, your body, or your favorite sports team. Each is a system made up of smaller, interconnected subsystems that interact with each other and affect how the others operate.

Testing as you fly requires a multilayered approach. Rocket scientists begin testing with the subcomponents—for example, the individual cameras that will form a rover's vision system, as well as the cables and connectors. Once the cameras are fully assembled, the vision system is tested again as a whole.

The reason for this approach is well summarized by a Sufi teaching: "You think that because you understand 'one' that you must therefore understand 'two' because one and one make two. But you forget that you must also understand 'and.'"[10] Components that otherwise function properly may refuse to play nice with each other after assembly. Put another way, systems may produce different effects than do the individual components standing alone.

These system-level effects can be disastrous. A medication may deliver terrific results on its own but prove lethal when it interacts with other drugs. Plug-ins on your website may work well in isolation but cause catastrophe as a system. Individually talented athletes may function horribly when assembled as a team.

We can call this problem Frankenstein's monster. Its limbs come from human bodies. But once the pieces are stitched together, the result is unhuman.

Consider the case of another monster's awakening. When Adolf Hitler came to power, the German constitution was one of the "most sophisticated" of its day.[11] It contained two seemingly harmless provisions. One provision allowed the German president to declare a state of emergency—a declaration that the parliament could cancel by a simple majority vote. The other allowed the president to dissolve the parliament and call for new elections. German parliaments had a habit of fragmenting and deadlocking, so this second provision was intended to check against that problem. Although they seemed benevolent when viewed in isolation, these two provisions turned malevolent in combination, producing what constitutional law scholar Kim Lane Scheppele calls a "Frankenstate."

In the early 1930s, President Paul von Hindenburg invoked his constitutional power to dissolve a hopelessly deadlocked parliament. Before elections were held to elect a new parliament, Hindenburg issued a state of emergency at the urging of Chancellor Hitler. The declaration suspended almost all civil liberties in Germany. Although the parliament had the constitutional power to override the emergency decree, there was no legislature in place to invoke that power.[12] Schutzstaffel (SS) and Sturmabteilung (SA) agents immediately began a widespread cleansing of all opponents of the Nazi cause. Using the state of emergency as a pretext, the Nazis began to consolidate control and establish a one-party dictatorship with Hitler at the helm. Without a single constitutional violation, one of the world's most horrific states was born.

A similar type of design flaw was also a possible cause of the Mars Polar Lander crash in 1999.[13] As the lander was descending toward the

Martian surface using its rocket motors, its three stowed legs popped into place 1,500 meters above the surface. Although we don't know for certain what happened, the lander may have misinterpreted the jolt of the leg deployment as safe touchdown on the surface. But the lander hadn't touched down. It was still descending. The computer prematurely cut the descent engines off and sent the lander on a fatal plunge.

The Mars Polar Lander team had tested the landing on the ground, including the deployment of the legs. When the team first ran the test, the electrical switch on the legs had been wired incorrectly and didn't produce a signal. Members of the team discovered the miswiring and reran the test. But because they were running behind schedule, they focused only on the actual touchdown. They skipped the leg deployment, which would have happened before the touchdown during flight. Although the test showed that the switches were wired correctly, the fatal flaw remained concealed in the leg deployment. NASA didn't retest the leg deployment phase with the proper wiring. The result was a smoking hole on the Martian surface.

As these examples show, failure to conduct systems-level testing can produce unpredictable consequences. When you make a last-minute change to a product and ship it out the door without retesting the whole thing, you're risking disaster. When you make a change to a section of a legal brief without examining how the change interacts with the whole, you're dancing with malpractice. When you subcontract the design of a major government program to sixty contractors but fail to test the combined system—as happened with healthcare.gov—you're courting catastrophe.[14]

In rocket science, there is another system that needs to be tested before takeoff. This system is far more unpredictable than the spacecraft itself. It panics. It forgets things. It tends to bump up against other objects or accidentally press the wrong button on the console. It can give in to fits of anger, develop a cold, or neglect important work to take in the cosmic scenery.

I am, of course, talking about the human beings on board.

The Right Stuff

The *right stuff* was the nickname given to the seven brave astronauts selected for NASA's first manned space mission, Mercury. Yet equally deserving of this title is another group of volunteers, whose names you've never heard.[15] NASA recruited these volunteers to take part in a series of tests on Earth to simulate flight conditions in space. In 1965, seventy-nine Air Force personnel, donning spacesuits, took rides on a space capsule attached to an impact sled. They rode the capsule "upside down and right side up, backward, forward, sideways, at 45-degree angles." Although a typical human will lose consciousness at about 5 g-forces—the *g* is short for the gravitational acceleration on the Earth's surface—the volunteers were subjected to g-forces peaking at a whopping 36.[16]

The goal of these experiments was to test as you fly—to subject unsuspecting Air Force personnel to the same types of shocks the astronauts would experience on the lunar journey. The volunteers damaged their eardrums and suffered compression injuries. One man ruptured his stomach after riding the capsule "with his rear end up in the air." Another was found to have an eye that was "off a little bit." Colonel John Paul Stapp, who ran the experiments, summed them up in a press release as follows: "At the cost of a few stiff necks, kinked backs, bruised elbows, and occasional profanity, the Apollo capsule has been made safe for the three astronauts who will have perils enough left over in the unknown hazards of the first flight to the moon."

The test-as-you-fly rule explains why we sent our closest cousins to space before we sent humans.[17] Because we had little idea about the effects of weightlessness on the human body, the first American in space was Ham the Chimp. Ham survived the flight, suffering only a bruised nose, and later died of natural causes (he is buried at the International Space Hall of Fame, where he was eulogized by Colonel Stapp).

Ham had been trained to perform basic tasks such as pulling levers, which he successfully replicated during his sixteen-minute flight. Although Ham's flight was a success, it bruised the fragile egos of the Mercury astronauts, who quickly realized that chimps were equally qualified to do their job. When President Kennedy's daughter, Caroline, met the astronaut John Glenn, the disappointed four-year-old reportedly asked, "Where's the monkey?"

We no longer send chimps to space or apply medieval torture techniques to Air Force volunteers. The methods have changed, but the underlying commitment to the test-as-you-fly rule remains. The day-to-day reality of an astronaut's life is vastly different from the glamour you see in Hollywood movies. Astronauts are workhorses, not space adventurers. They don't fly in space for a living. They train and prepare for spaceflight for a living. "I've been an astronaut for six years," explained Chris Hadfield, "I've been in space for eight days."[18]

The remainder is spent on preparation. By the time astronauts fly on their mission, they have flown the same route countless times on simulators.[19] For example, the mock-up of the space shuttle was outfitted just like the real thing, with identical controls and displays. The astronauts operated the space-shuttle simulator as they would operate the actual spacecraft, working through different segments of the mission, from launch to docking to landing. The monitors on the simulator displayed the same scenes the astronauts would see in flight, and hidden loudspeakers generated the same noises—including vibrations, pyrotechnic explosions, and gear deployment—they would hear in flight.

But there's one thing simulators can't do: generate microgravity. That's where the vomit comet comes into play.[20] It's the name for an airplane that flies in parabolas—kind of like a roller coaster—climbing and then diving to simulate weightlessness. At the top of each parabola, passengers experience about twenty-five seconds of microgravity. The airplane earned its name because these steep climbs and sharp dives tend to produce severe bouts of nausea among the passengers. Astronauts hop on board the vomit comet to practice moves like eating and drinking while floating in weightlessness.[21]

But twenty-five seconds isn't enough to practice more complex moves. For longer periods of weightlessness, astronauts dive into a giant indoor pool called the Neutral Buoyancy Lab. The buoyancy of water simulates the type of microgravity they'll experience in space.[22] "I really feel like a full-fledged astronaut in the pool," Hadfield writes, "I'm wearing a spacesuit, my breathing is assisted just as it is during a spacewalk." In the pool, which contains mock-ups of the International Space Station, astronauts practice making the same types of repairs they'll eventually conduct while floating in outer space (also called a spacewalk). They practice every step until it becomes second nature.[23] For Hadfield, achieving this level of familiarity meant spending 250 hours in the pool to get ready for a 6-hour spacewalk.[24]

Astronaut simulations are directed at NASA by a simulation supervisor—or SimSup for short—who leads a team of instructors.[25] One part of SimSup's job is to teach the astronauts the correct procedures for every segment of the mission. The other part of the job is far more grim: kill the astronauts.

The simulation team plays its own version of the kill-the-company exercise we encountered earlier—where corporate executives play the role of a competitor seeking to put the company out of business. The goal of the kill-the-astronaut exercise is similar. It's to push the astronauts to make the wrong moves in the simulator so they learn to make the right ones in space. When something goes astray in space, there's often no room for prolonged deliberation. Testing as you fly requires reducing the response time as close to instantaneous as possible. For space shuttle missions, this preparation meant activating roughly 6,800 malfunction scenarios, throwing every imaginable failure—computer crashes, engine troubles, and explosions—at the crew.[26] During the training of the Apollo astronauts, as author Robert Kurson explains, these simulations would run for days at a time. "The more catastrophic the better, until repetition began to groove instinct into all the participants, and dying helped the men learn to survive."[27]

In many ways, these simulations are tougher than the actual flight. They follow the old adage "The more you sweat in peace, the less you bleed in war." When Neil Armstrong first began walking on the lunar surface, he noted how the actual experience was "perhaps easier than the simulations at one-sixth g," referring to the reduced gravity on the Moon.[28] Sweating the small stuff on Earth ensured that the same stuff didn't make Armstrong bleed in space.

Repeated exposure to problems inoculates astronauts and boosts their confidence in their ability to defuse just about any issue. When physics throws curveballs at them, their training kicks in. After Hadfield returned to Earth from a successful mission, he was asked if things had gone as planned. "The truth is that nothing went as we'd planned," he responded, "but everything was within the scope of what we prepared for."[29]

Apollo astronaut Gene Cernan spoke in similar terms about his training. "If the [spacecraft] went somewhere we didn't like or the ground didn't like," he said, "I could flip a switch and I could control over 7.5 million pounds of rocket thrust [and] fly the thing to the Moon myself." Cernan was the commander of the Apollo 17 mission and the last person to leave footprints on the lunar surface. He continued: "I had practiced it and trained for it so many times, I almost dared her, I almost dared her to quit on me." After repeated practice, the astronaut and the spacecraft had fused into one. "Every breath she breathed," Cernan recalled, "I breathed with her."[30]

When the oxygen tank exploded on the Apollo 13 mission—literally taking the astronauts' breath away—their training kicked in. The movie *Apollo 13* displays a chaotic environment on the spacecraft and in mission control, with rocket scientists and astronauts scrambling to improvise solutions. Because the service module was damaged from the explosion, they had to figure out how to use the lunar module—intended only to shuttle two astronauts to the lunar surface—as a lifeboat for returning all three astronauts back to Earth.

But the reality was much calmer than its Hollywood depiction. Gene Kranz, the flight director for the mission, had conducted regular dress rehearsals to train mission controllers to solve complex problems in stressful situations.[31] In fact, a similar type of contingency requiring the astronauts to use the lunar module as a lifeboat had been simulated before. "No one had ever simulated exactly what happened," Apollo astronaut Ken

Mattingly explains, "but they had simulated the kind of stress that could be applied to the system and the people in it. They knew what their options were, and had some ideas already in place about where to go."[32]

This training strategy is useful far beyond rocket science. Consider, for example, oral arguments before the US Supreme Court. The highest judicial body in the land, the court hears less than one hundred cases every year, with a small number of the country's premier attorneys getting the privilege to present arguments before it.

I remember when I first walked into the courtroom as a visitor. The first thing I noticed wasn't the grandeur, the tall ceiling, or the marbled walls. No, it was how frighteningly close the attorney's lectern is to the mahogany bench where the nine Supreme Court justices sit. As attorneys present their arguments to the Court, they get interrupted by sharp, and often confrontational, questions from the justices. For every half hour of argument, a lawyer can expect an average of forty-five questions.[33] Questions often pour in before the attorneys even finish their first sentence. Given the short distance between the lectern and the bench, lawyers literally get blindsided by justices out of their vision.

Emotional appeals may work before a jury, but not before nine of the greatest legal minds in the country. The lawyers must be cool and collected while giving instantaneous responses to a barrage of questions. "You have to think not just how the answer to this question is going to work," explains frequent Supreme Court advocate Ted Olson, "but what that's going to mean for other yet unasked questions. And you don't want to please one Justice and alienate two others at the same time."

It takes a rocket-science mindset—and rocket-science preparation—to master this mental roller coaster. Before he became a judge, John Roberts—the current chief justice of the Supreme Court—was widely considered one of the best oral advocates to ever have appeared before the court. To prepare for arguments, Roberts would draft hundreds of questions he could conceivably receive from the judges. He would prepare answers for every single one, but he knew that simply writing the answers wouldn't be enough. On argument day, the questions would be thrown at him in random order from different judges. To bring the test closer to flight, he would "write the questions on flash cards, shuffle them, and test himself, so he'd be prepared to answer any question in any order."

When Roberts stepped up to the lectern to deliver his argument, he looked like a natural. Jonathan Franklin, a former colleague, recalls the effect: "He was able to take complicated points, distill them to their essence and respond with an absolute minimum of verbiage, and make it seem that his argument is so obviously correct that you have no choice but to agree with him." His delivery was so smooth that it seemed to unsuspecting observers Roberts had heard the questions before and knew exactly how to respond.

Another lawyer applied the same mindset to her athletic training. When she started competing, Amelia Boone was a lawyer at a major Chicago law firm. On a typical training day, Boone would go for a run in a wetsuit, dunking herself in and out of the icy waters of Lake Michigan, with the frigid winter wind whipping against her face.[34] Observers decked out in layers of thick winter clothing would assume these were the delirious actions of a masochist. But no, the Queen of Pain—as Boone has come to be known—was getting ready for the World's Toughest Mudder.

The World's Toughest Mudder makes a marathon look like a casual stroll. The race is run for twenty-four hours nonstop. Participants must fight off sleep while conquering roughly twenty of the "biggest, baddest" obstacles scattered across a five-mile course. It's survival of the fittest: Whoever completes the most laps wins.[35]

Some of the obstacles are in water, which can drop down to freezing temperatures. To prevent hypothermia, all participants run in wetsuits. While the runners are on land, the wetsuit helps preserve the body's warmth since body heat tends to dissipate over the grueling twenty-four hours.

When Boone first started training, she had little strength. She spent six months trying to do a single pull-up, but failed miserably. In her first race, she fell off all the obstacles. "I was really bad at that," Boone told herself after the race. "Let's try and get better." And get better, she did. She's now a four-time world champion and among the best obstacle racers in the world—period—not just in her gender category.

Boone's secret is the same as any self-respecting astronaut: test as you fly. Train in the same environment you'll experience on race day—while your competition trains from the comfort of a gym because it happens to be raining outside. "You don't race on a treadmill with Netflix in front of you," Boone says, "so you shouldn't be doing your training like that."

The rain, the snow, the dark, the cold, the wetsuit—they all beckon Boone. By the time the race rolls around, she has been desensitized to the brutal conditions awaiting her. She greets them with a smile that seems to say, "Nice to see you again. Let's dance."

In our lives, we don't do what Roberts and Boone do. We train in conditions that don't mimic reality. We practice a major speech in the comfort of our home, when we're fully rested and awake. We do mock job interviews in our sweatpants with a friend using a predetermined set of questions.

If we applied the test-as-you-fly rule, we would practice our speech in an unfamiliar setting, after downing a few espressos to give us the jitters. We would do mock interviews while wearing an uncomfortable suit, with a stranger ready to throw curveballs at us.

Businesses can also benefit from this principle. Corporate simulations, if they follow the test-as-you-fly rule, can "enhance an organization's ability to make high-stakes decisions," as three business school professors write in the *Harvard Business Review*.[36] For example, Morgan Stanley conducts drills to determine how to respond to various threats, including hackers and natural disasters. One aerospace firm holds dress rehearsals to determine how to respond to moves from their competitors, such as mergers or alliances. "By engaging in dress rehearsals," the researchers explain, "participants get to know each other's strengths or weaknesses, and informal roles become clear."

The test-as-you-fly rule, as we'll see in the next section, can also help everyone, including companies and comedians, run focus groups and gauge public opinion on their next product or brand-new joke.

The Rocket Science of Public Opinion

If Apple had violated the test-as-you-fly rule, the iPhone wouldn't have seen the light of day.

One of the most profitable consumer products in modern history, the iPhone was a flop in surveys conducted before it was released. When asked in a survey whether they "like the idea of having one portable device" to fulfill all their needs, only about 30 percent of Americans, Japanese, and Germans said yes.[37] They seemed to prefer carrying around a separate phone, a separate camera, and a separate music player instead of a single device that could perform all three functions. Echoing the survey results, then Microsoft CEO Steve Ballmer said, "There's no chance that the iPhone is going to get any significant market share. No chance."

The iPhone didn't prove the survey wrong. As author Derek Thompson explains, the survey accurately measured the participants' "indifference to a product they had never seen and did not understand." In other words, the survey had failed to follow the test-as-you-fly rule. Hypothetically thinking about the iPhone was nothing like seeing it in person. Once consumers saw the iPhone in an Apple Store—once they stepped into the brand and held the revolutionary new device in their hands—they couldn't let it go. Their indifference quickly morphed into desire.

There's a question that's commonly asked by companies to customers in pricing experiments: How much would you pay for this pair of shoes? Think about it. When was the last time someone asked you this question in real life? My guess is never. It's one thing for customers to say they'll buy a hypothetical shoe at a hypothetical price point. But it's another for them to reach into their wallet, pull out their hard-earned dollars, and hand over the money to a cashier. The shoe company is far better off building an actual prototype, placing it in an actual store, and selling it to an actual customer—in other words, testing as they fly.

One man understood this concept more than any other person did. If you've ever seen the results of a public-opinion poll, you've heard his name.

George Gallup was interested in finding an objective way to determine reader interest in newspapers.[38] He decided to write his PhD dissertation on the topic, appropriately titling it "An Objective Method for Determining Reader Interest in the Content of a Newspaper." For Gallup, the operative word was *objective*. He was deeply skeptical of subjective methods of determining reader interest, particularly the use of surveys and questionnaires. He believed—correctly, it turned out—that when it comes to reporting their own behavior, people tend to bend the truth. Readers would claim in surveys that they read the front page of the newspaper in full, but in reality, they would skip to the sports or the style section.

Put differently, these surveys failed the test-as-you-fly rule. Filling out a survey about reading a newspaper, and the actual act of reading a newspaper, are two different things. Gallup knew that for the test to work, it had to closely resemble the flight.

So what did Gallup do to remedy the problem? He sent a team of interviewers into people's homes to watch them read newspapers and mark each part of the paper as read or unread. Awkward? Yes. More accurate than surveys? Absolutely. "Almost without exception," Gallup wrote, "later questioning proved . . . preliminary statements [in surveys] false." Gallup's analog experiment was the precursor to modern digital tracking. If you think his approach was creepy, remember that Netflix knows exactly what you watch, when you watch it, and whether you stopped the last season of *House of Cards* before it ended. Netflix knows, as Gallup did, that observation is far more accurate than self-reporting.

Great comedians also think like rocket scientists and test their material before an actual audience to observe their reaction. They pop into comedy clubs unannounced to test their material in a low-stakes environment filled with strangers. For example, before hosting the Oscars in 2016, Chris Rock dropped by the Comedy Store, a comedy club in Los Angeles, to test his material.[39] Ricky Gervais and Jerry Seinfeld also visit tiny comedy clubs and adjust their jokes—or drop them altogether—in light of audience reaction.[40]

It's one thing to drop into random comedy clubs or watch people as they read newspapers. But it's something else entirely to ask people to let a stranger walk into their bathrooms and watch their children as they brush their teeth. The global design firm IDEO did exactly that after being tasked by Oral-B with designing a better toothbrush for children. Oral-B executives initially rolled their eyes at IDEO's unorthodox and mildly disturbing request. "It's not rocket science," the executives protested. "We're talking about kids brushing their teeth."[41]

It turns out that it *is* rocket science. Designing a great toothbrush, just like designing a great rocket, requires synergy between test and flight. Let's set aside the amusing image of an IDEO staffer busily taking notes while a five-year-old tries to concentrate on the already-challenging task of brushing teeth. Instead, let's focus on what IDEO uncovered. Before IDEO came along, manufacturers of children's toothbrushes assumed that children, who have smaller hands, needed smaller toothbrushes. So they took adult-sized toothbrushes and made them skinnier.

This approach sounds intuitive, but it completely missed the mark. IDEO's field research uncovered that children brush their teeth differently than how adults do it. Unlike adults, children grab the entire toothbrush with their fist. They lack the type of dexterity that adults have to move the toothbrush around with their fingers. Skinny toothbrushes make the children's job even harder, since the toothbrush tends to move around in their hands as they try to brush. As a result, what children needed were big, fat toothbrushes. Despite their initial skepticism of IDEO's approach, Oral-B executives went with IDEO's recommendations, producing a toothbrush that became the best seller in its category.

IDEO used the same strategy in redesigning the patient experience in a hospital. These institutions are supposed to nurse patients back to life, but most hospital rooms do the opposite. They are featureless, soulless white rooms lit by fluorescent lights.

When one health-care organization brought in IDEO to redesign the patient experience, the executives were probably expecting a stylish

PowerPoint presentation with new, creative designs for the hospital rooms. Instead, what they got was a mind-numbing six-minute video clip. The video showed nothing but the ceiling of a hospital room. "When you lie in a hospital bed all day," IDEO's chief creative officer Paul Bennett explained, "all you do is look at the roof, and it's a really shitty experience."[42]

What Bennett describes as "a blinding glimpse of the bleeding obvious" came after IDEO employees put themselves in a patient's shoes. An IDEO designer checked in to the hospital as a patient and lay in an actual patient bed for hours, getting wheeled around, staring at the ceiling tiles, and capturing the abysmal experience on a video camera. That six-minute clip of the dull tiles was a small glimpse of the overall patient journey—a "mix of boredom and anxiety from feeling lost, uninformed, and out of control," as IDEO's CEO Tim Brown said.[43]

Six minutes of footage was sufficient for the hospital employees to spring into action. They decorated the ceilings, put up whiteboards for visitors to leave messages to the patients, and transformed the style and color of the patient rooms to make them more personal. They also put rearview mirrors on hospital stretchers to allow patients to see and connect with the doctors and nurses wheeling them around. IDEO's presentation ultimately kick-started a broader discussion to improve the overall patient experience so that patients were "treated less like objects to be positioned and allocated, and more like people in stress and pain," Brown explained.[44]

As these examples show, instead of creating artificial testing environments disconnected from reality, we're better off observing customer behavior in real life. If you want to design a better newspaper, watch people read the paper. If you want to design a better kid's toothbrush, watch kids brush their teeth. If you want to see if people will love the iPhone, put an iPhone in their hands. "If you want to improve a piece of software," as IDEO's founder David Kelley explains, "all you have to do is watch people using it and see when they grimace."[45]

This approach provides a vast improvement over subjective self-reporting in artificial conditions. But it doesn't completely eliminate the distance between the test and the flight. Observing people, it turns out, tends to affect how they behave.

The Observer Effect

The observer effect is among the most misunderstood concepts in science. It has given rise to pseudoscientific claims that the conscious mind can magically alter reality and make a spoon move across the dinner table. But the scientific concept is simple. By observing a phenomenon, you can affect that phenomenon. Let me explain.

I started wearing glasses when I became a professor. But true to the absent-minded professor stereotype, I tend to misplace them. If I'm looking for my glasses in a dark room, I do what everyone else does: turn on the light. The act of turning on the light sends a flood of photons toward my pair of glasses, which reflect off the glasses and into my eyes.

But now assume that, instead of my glasses, I'm trying to find an electron. To observe an electron, I do the same thing and send some photons in its direction. My glasses are relatively big objects, so when photons collide with them, the glasses don't move. But when photons collide with an electron, they displace the electron. You can also think of it as a coin lodged between the cushions of a couch.[46] The very act of trying to grab the coin pushes it further out of reach.

The act of observation disturbs humans in a different way. When people know they're being observed, they behave differently.

Let's suppose you're in a test audience for a new television show. Watching the show as part of a focus group is a different experience from watching it in your living room. The test isn't identical to the real flight. In a focus group, you may find numerous flaws in a show—since you're being observed by people who asked you to critically evaluate it—even though the same show would have proved binge-worthy in your living room.

For example, the TV show *Seinfeld* performed abysmally before test audiences.[47] When creating the show's premise, the producers asked a question we encountered in an earlier chapter: "What if we did the opposite of what everyone else is doing?" At the time, the sitcom playbook was set in stone. A group of characters would run into problems, resolve those problems, learn something from the experience, and hug each other.

From the get-go, *Seinfeld* producers were clear on their mission. They would flip the script. There would be no hugging. There would be no learning. The characters on *Seinfeld* would draw laughs from constantly repeating their mistakes and overlooking their own faults. In case there was any confusion, the writers wore jackets that said No Hugging, No Learning. But test audiences, accustomed to the standard sitcom playbook, were expecting lots of hugging and learning. As a result, *Seinfeld* was a spectacular failure in focus groups. Yet the show went on to become one of the most popular sitcoms of all time.

The observer effect is often an unconscious process. Even when we assume we're not affecting the participants—even when we're careful to not dislodge that coin in the couch cushion—we might be cuing them in subtle but significant ways.

Consider Clever Hans the horse.[48] Hans was the closest thing a horse could come to a rocket scientist. He became a worldwide sensation for his ability to perform basic math. Its owner, Wilhelm von Osten, would ask the audience for a math problem. Someone would shout out, "What is six plus four?" and Hans would tap his hoof ten times. His ability went beyond addition. He could subtract, multiply, and even divide. People suspected fraud, but independent investigators found no foul play.

It was a young psychology student named Oskar Pfungst who figured out what was really going on. Hans could find the right answer only if he could see the human questioner. His mathematical genius disappeared if he was wearing blinders or otherwise couldn't see the human intermediary. In the end, it was the human questioners who were unwittingly providing cues to the horse. As Stuart Firestein writes, "People would tense the muscles of their body and face at the beginning of Hans's answer and release the tension when he arrived at the correct hoof tap." Remarkably, even after Pfungst discovered Hans's secret, he couldn't prevent himself from unconsciously giving cues to the horse. As long as he knew the answer, Pfungst's demeanor would involuntarily shift when Hans arrived at the correct hoof tap.

The distortions introduced by the observer effect are significant. The effect can fool you into believing that a hit show will be a flop or that a horse is a mathematical genius.

One way to mitigate this effect is to put blinders on both the human questioner and the horse by conducting what's called a double-blind study. For example, in drug trials, both the participants in the study and the scientists running it are kept in the dark—hence *double*-blind—about whether the participants are getting the actual drug or the fake one, called a placebo. If the methods aren't double-blind, scientists may insert their hopes and prejudices into the study, treating the participants differently or unconsciously cuing them like Hans's human questioners.

You can also take a cue from the best-selling author Tim Ferriss.[49] Most authors, in picking the title and cover design for their book, simply go with their gut or, at best, consult a few friends. The more astute ones run a survey of their audience. But Ferriss took this analysis to the rocket-science level with his first book.

To select a title, Ferriss applied the test-as-you-fly principle. He bought domain names for roughly a dozen book titles and ran a Google AdWords campaign to test click-through rates. When a user typed certain keywords into a Google search related to the book's content, an ad would pop up with the book title and subtitle that pointed to a dummy web page for a book that didn't yet exist. Google would automatically randomize and mix and match the book titles and subtitles that were displayed to the user, allowing an objective analysis of popularity. Within a week, it became apparent that *The 4-Hour Workweek* title attracted by far the most attention. Ferriss took the data to his publisher, which didn't need much convincing that the title was the right one.

But Ferriss didn't stop there. To choose a cover for his book, he went to a bookstore with alternative cover designs in hand. He picked up a book from the new-arrivals section and wrapped it in one of his covers. He then sat back and watched how many times the book was picked up by unsuspecting customers, repeating the exercise for each version of the cover for thirty minutes at a time until he settled on a winner.

One final piece of the testing puzzle is often overlooked. Perfectly planned tests can spit out perfectly incorrect results if the testing instrument itself is flawed.

Multiple Testers

The irony was hard to escape. A space telescope built to produce distortion-free images was spitting out distorted images.[50] The Hubble Space Telescope was launched in 1990 with the promise of taking detailed, high-resolution images of the cosmos—ten times sharper than what earthbound telescopes could produce. The size of a school bus, the telescope would hover above the Earth free of the distortions introduced by the atmosphere, providing the clearest view humanity has ever had into the cosmos.

But the first set of images that came down from the Hubble weren't nearly as clear as astronomers had hoped. The $1.5 billion telescope was suffering from myopia and sending fuzzy photos back to Earth.

It turned out the primary mirror of the telescope had been ground to the wrong shape because the testing device used to ensure correct grinding hadn't been set up properly. One of the lenses on the testing device—called the reflective null connector—was out of position by 1.3 millimeters (or 0.05 inches). This positioning produced a flaw on the mirror that was one-fiftieth the thickness of a piece of paper. That may seem like a minor flaw, but when it comes to sensitive instruments, millimeters can be mountains. Over the course of five years of grinding and polishing, the mirror had been very precisely ground to the wrong shape.

The commission that was convened to investigate the mirror fiasco criticized the use of a single instrument to test the mirror. Because of cost and scheduling concerns, the team had dismissed the need for independent testing by a second instrument.

Here's the moral of the Hubble story. If you're going to rely on a single instrument for testing—and put all your proverbial eggs in the same basket—you must test the basket to make sure it won't cave. But this wasn't done in the case of the Hubble. No one had tested the testing device to ensure a correct setup and accurate spacing of its lenses.

Thankfully, there was a contingency in place, and the telescope could be serviced in space. Astronauts did what you do when your vision is blurry. They put eyeglasses on the Hubble. Because the flaw on the Hubble's primary mirror was perfectly wrong, the perfect prescription could right the wrong. In a servicing mission in 1993, astronauts outfitted the Hubble with spectacles, restoring the telescope to its promised glory and returning it to its mission of producing dazzling images that now decorate computer backgrounds across the globe.

Consider another example from outside rocket science.[51] The Facebook website was originally designed in 2006, when the "web was a lot more text heavy," as Julie Zhuo, Facebook's vice president of product design, told me. With the rise of camera phones, the company wanted to create a more visual experience. After six months of work, the Facebook team created a modern, cutting-edge website. They tested the new website internally, and it worked beautifully. They hit publish and waited for all the praise to pour in.

But the company was in for a rude awakening. The metrics showed the redesign was a colossal failure. "People started using Facebook less. People were commenting and engaging with other people less," Zhuo explained to me.

It took the Facebook team several months of fieldwork to figure out what had gone wrong. The team had tested the new website using high-tech computers in Facebook offices. But the vast majority of Facebook users lacked access to top-of-the-line equipment. They were accessing the website with old computers that didn't support all the fancy imagery that came with the redesign. Put differently, for most Facebook users, the flight was far different from the test. Only when the Facebook team switched out its testing instruments—and used low-tech instead of high-tech equipment—did the group create a redesign that worked for its user base.

These examples hold important lessons for us all. Treat your testing instruments like your investments and diversify them. If you're building a website, test it using different browsers and different computers. If you're designing a children's toothbrush, watch many children brush their teeth—lest you get the one miracle child who uses a toothbrush like an adult. If you're deciding which job offer to take, consult multiple calibration targets. One person's opinion might provide only a fuzzy perspective. It's only through independent validation and multiple testing sources that you get closer to twenty-twenty vision.

....

WHETHER IT'S LAUNCHING a rocket, training for a sporting event, arguing before the Supreme Court, or designing a telescope, the underlying principle is the same. Test as you fly—subject yourself to the same conditions you'll experience during the flight—and you'll soon begin to soar.

Visit **ozanvarol.com/rocket** to find worksheets, challenges, and exercises to help you implement the strategies discussed in this chapter.

STAGE THREE
ACHIEVE

In this last section of the book, you'll learn why the final ingredients for unlocking your full potential include both failure and success.

8

NOTHING SUCCEEDS
LIKE FAILURE

How to Transform Failure into Triumph

Man errs as long as he strives.

—GOETHE

I N THE EARLY stages of development, rockets tend to blow up, drift off
course, and otherwise explode. The rockets that were launched as precursors to the Moon landing were no exception. Problems occurred in
virtually every mission.

In December 1957, two months after the Soviet satellite *Sputnik* had become the first in Earth orbit, Americans attempted to even the score.[1] The
rocket, called *Vanguard*, launched about four feet above the pad, hesitated,
and then sank back down, exploding on national television and earning
itself nicknames like *Flopnik*, *Kaputnik*, and *Stayputnik*.[2] The Soviets were
quick to rub salt in the Americans' cosmic wound. They inquired whether
the United States was interested in receiving foreign aid earmarked for
"undeveloped countries."

In August 1959, the unmanned rocket *Little Joe 1* got a little too excited. Because of an electrical problem, it decided to launch itself half an
hour before schedule, as NASA personnel watched dumbfounded.[3] It

crashed after flying for twenty seconds. In November 1960, the launch of the *Mercury-Redstone* rocket became known as the "four-inch flight." The rocket lifted just four inches off the ground before settling back down on the pad.[4]

Numerous mishaps occurred in manned missions as well. To cite one memorable example, a problem on Gemini 8 nearly claimed the life of Neil Armstrong three years before he stepped foot on the Moon.[5] Gemini 8 was a complex mission that would mark the first time two spacecraft docked in orbit. A radio-controlled target vehicle, called the *Agena*, would be launched into orbit first, followed by *Gemini 8*, which would rendezvous and dock with the *Agena*.

The successful docking was followed by panic. Long before the movie *Apollo 13* made the line famous, astronaut Dave Scott radioed to Houston, "We have serious problems here." Gemini 8 had started spinning wildly— more than one revolution per second—blurring the astronauts' vision and threatening vertigo and loss of consciousness. As the spacecraft continued to spin out of control, a cool and collected Armstrong jettisoned the *Agena*, switched to manual controls, and fired the opposite thrusters to slow down the spin.

....

THE "FAIL FAST, fail often, fail forward" mantra is all the rage in Silicon Valley. Failure is viewed as inspirational fodder, a rite of passage, a secret handshake shared by the insiders. Countless business books instruct entrepreneurs to embrace failure and flaunt it as a badge of honor. There are conferences, such as FailCon, dedicated to celebrating failure and FuckUp Nights, where thousands have gathered in more than eighty-five countries to toast their failures.[6] There are funerals for failed start-ups, complete with bagpipes, DJs, sponsorships by liquor companies, and slogans like "Putting the Fun in Funeral."[7]

Most rocket scientists would bristle at this cavalier attitude toward failure. In rocket science, failure can mean the loss of human life. Failure can also cost taxpayers hundreds of millions of dollars. Failure means that decades of work goes up in smoke—literally and figuratively. No one celebrated the numerous explosions and mishaps that occurred during the race to the Moon. They were embarrassing. They were catastrophic. And they weren't taken lightly.

In this chapter, I'll use a rocket-science framework to explain why it's as dangerous to celebrate failure as it is to demonize it. Rocket scientists apply a more balanced approach to failure. They don't celebrate it; nor do they let it get in their way.

In the first and second parts of this book ("Launch" and "Accelerate"), we explored how to ignite, refine, and test breakthrough ideas. Pursuing audacious ideas means daring greatly, and daring greatly means some of those ideas are going to fail when they collide with reality. So we begin this last stage of the book—"Achieve"—with failure.

You'll learn why most of us think about failure the wrong way and how we can redefine our relationship with it. I'll reveal how elite companies build failure into their business models and create an environment where employees willingly reveal their mistakes, instead of concealing them. I'll share with you one of the biggest misconceptions about rocket science that appears in a Hollywood blockbuster and what the development of Viagra can teach you about failure. You'll walk away from the chapter with science-backed ways to fail gracefully and create the right conditions for learning from failure.

Too Afraid to Fail

We're wired to fear failure. Centuries ago, if we didn't fear failure, we became prey to a ravenous grizzly bear. Growing up, failure got us into the principal's office. Failure meant getting grounded or getting our allowance cut. Failure meant dropping out of college or not getting our dream job.

There's no denying it; failure sucks. In most aspects of life, there are no participation trophies. When we fail a class, go bankrupt, or lose our job, we're in no mood to celebrate. We feel worthless and weak. Unlike the high of success, which quickly dissipates, the sting of failure lingers—sometimes for a lifetime.

To ward off the bogeyman of failure, we keep a safe distance from it. We stay off the edges, avoid healthy risks, and play it safe. If we aren't guaranteed to win, we assume the game isn't worth playing.

This natural tendency to avoid failure is a recipe for failing. Behind every rocket unlaunched, every canvas unpainted, every goal unattempted, every book unwritten, and every song unsung is the looming fear of failure.

Thinking like a rocket scientist requires redefining our troubled relationship with failure. It also requires correcting one of the greatest misconceptions about rocket science popularized by a Hollywood blockbuster.

Failure Is an Option

In *Apollo 13*, there's a scene where a group of rocket scientists are gathered in a room after learning that the spacecraft suffered an oxygen tank explosion on its voyage to the Moon. The spacecraft's power is dangerously low, and the astronauts' days are numbered. The scientists in mission control must figure out a way to get them back before their power runs out. "We never lost an American in space. We're sure as hell not going to lose one on my watch," roars Gene Kranz, the flight director, before adding the punch line: "Failure is not an option." Kranz later wrote an autobiography with the same title, where he described the tagline as "a creed that we all lived by" in mission control.[8] NASA gift shops quickly capitalized on the credo and began selling T-shirts emblazoned with the words Failure Is Not an Option.

The mantra makes sense when you've got human lives at stake. But as a descriptor for how rocket science works, it's misleading. There's no such thing as a zero-risk rocket launch. You still have to compete with physics. You can plan for some mishaps, but the cosmic banana peel is always around the corner. Accidents are inevitable when you're creating a controlled explosion in a machine as complex as a rocket.

If failure weren't an option, we never would have dipped our toes into the cosmic ocean. Doing anything groundbreaking requires taking risks, and taking risks means you're going to fail—at least some of the time. "There's a silly notion that failure's not an option at NASA," Elon Musk says. "Failure is an option here [at SpaceX]. If things are not failing, you are not innovating enough."[9] It's only when we reach into the unknown and explore ever-greater heights—and in so doing, break things—that we move forward.

The same is true for scientists working in a lab. For them, without the ability to be wrong, they could never be right. Some of their experiments succeed and others don't. If things don't work as planned, it's a hypothesis proven wrong. They can tweak the hypothesis, try a different approach, or abandon it altogether.

British inventor James Dyson described the inventor's life as "one of failure."[10] It took Dyson fifteen years and 5,126 prototypes to get his revolutionary bagless vacuum to work. Several of Einstein's attempts to devise a proof for $E = mc^2$ failed.[11] In some fields—for example, pharmaceutical drug development—the average failure rate is over 90 percent. If these scientists lived by the "failure is not an option" mantra, the self-loathing, the shame, and the embarrassment would all cripple them.

A moratorium on failure is a moratorium on progress.

If you're in the business of taking moonshots—if you're going to experiment with bold ideas—you're going to miss more often than you connect. "Experiments are by their very nature prone to failure," Jeff Bezos explained. "But a few big successes compensate for dozens and dozens of things that didn't work."[12]

Remember the Amazon Fire phone? The company lost $170 million over that misfire.[13] Or Google Glass, designed by X, Google's moonshot factory?[14] The Glass was supposed to be the next best thing after the smartphone, but it flopped. It's one thing to carry a smartphone in your pocket, consumers thought, and something else to attach one to your cornea. This was one piece of hardware that was decidedly uncool to sport. People wearing it were dubbed "glassholes."

These failures are built into X's business model. To X, killing projects is a "normal part of doing business," as the company's head Astro Teller puts it. It's not unusual for X to kill over a hundred ideas in a single year.[15] "Because X is premised on the idea of pursuing highly risky proj-ects," Kathy Cooper of X explains, "there's just an understanding that a lot of them aren't going to work. So it's not seen as surprising or the fault of anyone if something doesn't work."[16] By normalizing failure, X makes moonshot thinking the path of least resistance.

Not everyone can afford to swing and miss to the tune of $170 million, as Amazon did with the Fire. The size of your investment may differ dra-matically, but the underlying principle remains the same: Treating failure as an option is the key to originality. "When it comes to idea generation," Adam Grant writes in *Originals*, "quantity is the most predictable path to quality."[17] Shakespeare, for example, is known for a small number of his classics, but in the span of two decades, he penned 37 plays and 154 sonnets, some of which have been "consistently slammed for unpolished prose and incomplete plot and character development."[18] Pablo Picasso produced 1,800 paintings, 1,200 sculptures, 2,800 ceramics, and 12,000 drawings—only a fraction of which are noteworthy.[19] Just a handful of Einstein's hundreds of publications had real impact.[20] Tom Hanks, one of my all-time favorite actors, admits, "I've made an awful lot of movies that didn't make any sense, and didn't make any money."[21]

But when we judge the greatness of these individuals, we don't focus on their troughs. We focus on the peaks. We remember the Kindle, not the Fire. We remember Gmail, not the Glass. We remember *Apollo 13*, not *The Man with One Red Shoe*.

It's one thing to acknowledge that failure is an option. But it's some-thing else entirely to celebrate it. To take the sting and shame out of failure, Silicon Valley overcorrected. The pendulum swung too far in the other direction.

The Problem with Fail Fast

The fail-fast mantra has no place in rocket science. When each failure is horrifically expensive—in terms of both money and human lives—we can't rush to the launch pad with a crappy rocket and fail as fast as possible.

Even outside rocket science, the fail-fast refrain is misguided. When entrepreneurs are too busy failing fast and celebrating it, they stop learning from their mistakes. The clinking of champagne glasses mutes the feedback they might otherwise receive from failure. Failing fast, in other words, doesn't magically produce success. When we fail, we're often none the wiser.

Consider a study of nearly nine thousand American entrepreneurs who founded companies between 1986 and 2000. The study compared the success rates—defined as taking a company public—of first-time founders and founders who had previously failed in business. You might expect that the experienced founders—having launched a business before and presumably learned from their failure—would be much more likely to succeed than those who had never started a business before. But that's not what the study found. The success rate of first-time entrepreneurs was nearly equal to the success rate of entrepreneurs who had previously failed in business.[22]

Another study is also on point. Researchers examined 6,500 cardiac procedures by seventy-one surgeons over a ten-year period. They found that the surgeons who botched a procedure performed *worse* on later procedures.[23] The results suggest that the surgeons not only failed to learn from their mistakes but also ended up reinforcing bad habits.

What explains these counterintuitive results?

When we fail, we often conceal it, distort it, or deny it. We make the facts fit our self-serving theory rather than adjust the theory to fit the facts. We attribute our failure to factors beyond our control. In our own failures, we overestimate the role of bad luck ("Better luck next time"). We blame the failure on someone else ("She got the job because the boss likes her more"). We come up with a few superficial reasons for why things went south ("If only we had more cash reserves"). But personal culpability seldom makes the list.

"What's a little white lie?" you might ask. After all, putting a positive spin on failure can help us save face. But here's the problem: If we don't acknowledge we failed—if we avoid a true reckoning—we can't learn anything. In fact, failure can make things worse if we get the wrong messages from it. When we attribute our failures to external factors—the regulators, the customers, the competitors—we have no reason to change course. We throw good money after bad, double down on the same strategy, and hope the wind blows in a better direction.

Here's what most people get wrong about persistence. Persistence doesn't mean repeatedly doing what's failing. Remember the old adage about the futility of doing the same thing over and over again and expecting different results? The goal isn't to fail fast. It's to learn fast. We should be celebrating the lessons from failure—not failure itself.

Learn Fast, Not Fail Fast

The hardest part of getting to Mars is clearing a hurdle right here on Earth. NASA doesn't build and operate Martian spacecraft all on its own.[24] When it's planning a new mission, it makes a formal announcement that describes, in general terms, the spacecraft that NASA intends to send and the science it expects the spacecraft to conduct. The announcement solicits proposals from anyone interested in sending scientific instruments to space. The number of great ideas far exceeds the funding, so NASA uses a Darwinian process to select only the strongest proposal. All others fail. This competitive system is as it should be: A cheap mission to Mars costs American taxpayers half a billion dollars.

My former supervisor, Steve Squyres, began writing proposals in 1987 to lead a Mars mission.[25] Over the ensuing ten years, every single one of his ideas was shot down. "It was bitterly disappointing when you put years of effort and hundreds of thousands of dollars into writing a proposal," Squyres recalls. But he doesn't accuse NASA of failing to see the genius in his proposals. Rather, he places the blame squarely on himself. "The early [proposals] weren't good enough," Squyres admits. "They didn't deserve to be selected."

There are two responses to negative feedback from a credible source: Deny it or accept it. Every great scientist chooses the latter, and Squyres did the same. Each proposal he submitted to NASA was better than the one that came before it.

After ten years of learning, tweaking, and improving, Squyres's proposal—which would eventually become the 2003 Mars Exploration Rovers mission—was finally selected in 1997. But selection didn't guarantee flight. The mission was scrapped and brought back to life three times—most recently, after the crash of the Mars Polar Lander, which, as described in earlier chapters, used the same landing mechanism our group was planning to use. The mission was salvaged by two questions that reframed the problem: What if we used airbags instead of a three-legged lander? And what if we sent two rovers instead of one?

After we doubled up on rovers—named *Spirit* and *Opportunity*—and got our launch ticket back, malfunctions occurred about every month. During testing, the parachutes displayed a problem called squidding. For unknown reasons, they would pulsate like a squid, opening up and flapping closed repeatedly—a problem not seen in a parachute like ours in thirty years.[26] One of the cameras on board the rovers developed an inexplicable "speckling" problem that overwhelmed the images with static.[27] Two months before launch, we blew the fuse on *Spirit*.

In late June 2003, I flew down to Florida for *Opportunity*'s launch. In advance of the launch, we got together on Cocoa Beach for a private team meeting with no agenda, our heads gazing skyward toward Destination Mars. As we were popping champagne corks to commemorate the occasion, we learned that the cork on our rocket had popped too.[28] The cork, which provides thermal insulation to the rocket, wouldn't stick and kept peeling off. Our launch was delayed for several days as we scrambled to find a solution. We came dangerously close to the end of our allotted launch window until someone came up with the ingenious idea of using a resilient superglue called red RTV, available at Home Depot. With red RTV to the rescue, we took off for the red planet.

Each failure proved to be an invaluable learning opportunity. Each failure revealed a flaw that required correction. Each failure was followed by progress toward the ultimate goal. Although these failures took their toll on us, we couldn't have landed safely on Mars without them.

These failures are what business school professor Sim Sitkin calls "intelligent failures." They happen when you're exploring the edges, solving problems that haven't been solved, and building things that may not work.

We often speak of intelligent failures as losses. "I lost five years of my life." "We lost millions of dollars." But these are losses only if you call them that. You can also frame them as investments. Failure is data—and it's often data you can't find in a self-help book. Intelligent failures, if you pay them proper attention, can be the best teachers.

These errors can have staying power that lessons from success often lack. Intelligent failures can produce a sense of urgency for change and produce the shock necessary to unlearn what we know. "Give me a fruitful error any time, full of seeds, bursting with its own corrections," Vilfredo Pareto wrote, "You can keep your sterile truth for yourself."[29]

Thomas Edison recounted the story of a conversation with an associate who lamented that after thousands of experiments, he and Edison had failed to discover anything. "I cheerily assured him that we had learned something," Edison recalled. "For we had learned for a certainty that the thing couldn't be done that way, and that we would have to try some other way."[30]

Learning can also take the stigma out of failure. "The best thing for being sad," the author T. H. White wrote, "is to learn something. That's the only thing that never fails. You may grow old and trembling in your anatomies, you may lie awake at night listening to the disorder of your veins, you may miss your only love, you may see the world about you devastated by evil lunatics, or know your honour trampled in the sewers of baser minds. There is only one thing for it then—to learn. Learn why the world wags and what wags it."[31]

Without opportunities to learn why the world wags and what wags it, failure has no upside. But if you've learned something—if this failure means you're more likely to succeed when you try again—failure won't hit you as hard. Learning takes despair and turns it into excitement. With a growth mindset, you can maintain forward momentum even as the explosions pile up, the work gets hard, and the obstacles begin to appear insurmountable. As Malcolm Forbes, the founder of *Forbes* magazine, put it, "Failure is success if we learn from it."

Squyres's failed proposals for Mars missions are still sitting on his desk. "I can look at those old proposals," he says, "and I can look at the things we did wrong, and I can look at the lessons that we learned and how we made things better, and I can see why on our fourth try we finally got selected."

Just a few years after our rovers set sail for the red planet, it would take another group of rocket scientists four attempts to get it right.

The Opening and the Finale

The third time's the charm.[32]

In August 2008, this is what SpaceX employees were telling themselves as they awaited the third launch of Falcon 1, the company's very first rocket. At the time, outside observers were already busy drafting the obituary of what they thought was Musk's vanity project. When Musk started SpaceX, he invested a hundred million dollars of his own money in the company—enough for three launches.

The first two failed.

Falcon 1's maiden flight in 2006 lasted for all of thirty seconds. A fuel leak caused an unexpected fire in the engine, shutting it off and sending the rocket plummeting into the Pacific. "The first launch failure was heartbreaking," SpaceX executive Hans Koenigsmann recalls. "We learned a lot of things we did wrong, and learning sometimes hurts." The leak was blamed on corrosion around an aluminum nut that secured the fuel line. To correct the problem, the company replaced the aluminum fasteners with stainless steel ones that were more reliable and had the benefit of being cheaper.

The Falcon 1 was back on the launch pad a year later in 2007 for a second attempt. This flight got farther—clocking in at 7.5 minutes—but also failed to reach orbit after fuel stopped flowing into the engine. The failure "didn't feel anywhere as harsh as the first time," Koenigsmann says. "The vehicle actually flew very far, and then didn't make orbit, but at least it flew out of sight." Despite the ultimate failure, most mission objectives were met: Falcon 1 could launch and reach space. The anomalies that caused the trouble were quickly diagnosed and fixed.

The third attempt came a year later. Although 2008 was a bad year for many people, Musk says it was the worst of his life. His electric car company, Tesla, was flirting with bankruptcy, the world had spun into a financial crisis, and Musk had just gotten divorced. He was borrowing money from his friends to pay rent. He had plunged much of his fortune into SpaceX, and the two Falcon 1 failures had eaten into his investment. What was left of it was sitting on the launch pad awaiting a perilous flight.

On the third attempt, Falcon 1 rumbled to life and took off carrying three satellites and the ashes of James Doohan, the actor who played Scotty in *Star Trek: The Original Series* (think "I'm givin' her all she's got, Captain!"). It soared into the sky, executing a perfect flight of its first stage (recall that rockets are built in stages stacked on top of each other). After the first stage took the spacecraft into space, it was time for stage separation—the critical point in the flight where the rocket's first stage detaches and falls away after running out of fuel. That's when the smaller, second stage kicks in to take the spacecraft into orbit. The stages separated as scheduled, but the first stage didn't stop. It fired again and bumped into the second. "We rear-ended ourselves," SpaceX's president, Gwynne Shotwell, recalls. "It was almost Monty Pythonesque."

The problem was missed during testing because SpaceX had failed to follow the test-as-you-fly principle. The engine pressure that resulted in the unexpected boost of thrust was below the ambient pressure in SpaceX's ground testing facility, so it had barely registered. But in the vacuum of space, the same pressure produced enough of a kick to cause a catastrophic collision.

For SpaceX, this failure was strike three. Hundreds of shell-shocked SpaceX employees, who had been working seventy- to eighty-plus-hour weeks for six years, awaited word from their boss at SpaceX's factory in

Hawthorne, California. "The mood in the building hung thick with despair," recalls former SpaceX employee Dolly Singh. Musk emerged from the control room, where he was commanding the mission along with senior engineers. He walked past the press to address his troops, who had just lost their third consecutive major battle.

Musk told them that they knew the project was going to be hard. He reminded them that what they were trying to do was, after all, rocket science. The company's rockets had reached space, accomplishing what major countries had failed to accomplish. Then came the surprise: Musk announced he had secured an investment that would get SpaceX two more launches. This wasn't the end. As Shane Snow describes it, Musk told his troops they would "learn what had happened tonight and they would use that knowledge to make a better rocket. And they would use that better rocket to make even better rockets. And those rockets would one day take man to Mars."[33]

It was time to get back to work. "Within moments," Singh recalls, "the energy of the building went from despair and defeat to a massive buzz of determination as people began to focus on moving forward instead of looking back." The likely culprit for the failure was identified within a matter of hours. "When I saw the video, it was like, 'OK. We can figure it out,'" Shotwell explains. The solution was simple: introduce a longer delay before stage separation to prevent a collision. "Between the third and the fourth flight we changed one number, nothing else," Koenigsmann says.

In less than two months, SpaceX was back on the launch pad. "Everything hinged on that launch," recalls Adeo Ressi, Musk's college friend. "Elon had lost all his money, but this was more than his fortune at stake—it was his credibility." If the fourth launch failed, "it would have been over. We're talking Harvard Business School case study—rich guy who goes into the rocket business and loses it all."

But the rocket didn't fail. On September 28, 2008, SpaceX's Falcon 1 launched out of the atmosphere and into the record books, becoming the world's first privately built spacecraft to reach Earth's orbit.

When SpaceX survived its baptism by fire on its fourth try, everyone took notice—particularly the bureaucrats at NASA looking to sustain the

American space program after the expected retirement of the space shuttle in 2010. In December 2008, three months after Falcon 1's successful voyage, NASA handed SpaceX a lifeline in the form of a $1.6 billion contract for resupply missions to the International Space Station. When NASA officials called to give him the good news, an otherwise stern Musk broke out of character and screamed, "I love you guys!" For SpaceX, Christmas had come early.

To paraphrase F. Scott Fitzgerald, there's a difference between a single failure and final defeat.[34] A single failure, as SpaceX's story illustrates, can be the beginning, not the end. Many outside observers called the three Falcon 1 crashes failures—mistakes committed by a team of amateurs led by a rich kid playing with expensive toys. But labeling these crashes failures was like calling a tennis match before it's over. "I've come from behind too often," the great tennis champion Andre Agassi writes, "and had too many opponents come roaring back against me, to think that's a good idea."[35]

The opening doesn't have to be grand, as long as the finale is.

Time morphs how we view events. Something that looks like a failure in the short term changes when we zoom out and put on a broader lens. Pixar's former president Ed Catmull calls the initial ideas behind the studio's blockbuster animation films "ugly babies." All their films start out "awkward and unformed, vulnerable and incomplete."[36] But if the game doesn't end until the film is released, an early version gone wrong isn't a catastrophe. It's a momentary blip. A temporary glitch. A problem to be solved.

Breakthroughs are often evolutionary, not revolutionary. Take a look at any scientific discovery, and you'll find there is no magical it. No single aha moment. Science weaves from failure to failure, with each version better than the one that came before. From a scientific perspective, failure isn't a roadblock. It's a portal to progress.

We embodied this mindset as children. When we learned how to walk, we didn't get it right on the first try. No one told us, "You'd better think hard about how you take that very first step because you get one step and that's it." We repeatedly fell. With each fall, our bodies learned what to do and what not to do. By learning not to fall, we learned how to walk.

Nothing comes to existence perfectly formed. Rome, as the saying goes, wasn't built in a day. The Apollo 11 spacecraft that put Armstrong and Aldrin on the Moon didn't just spring out of the factory. It took numerous iterations—through the Mercury, Gemini, and earlier Apollo missions—to get it right.

For scientists, each iteration is progress. If we get a glimpse into the dark room, that's a contribution. If we don't find what we thought we'd find, that's a contribution. If we change a single unknown unknown to a known unknown, that's a contribution. If we ask a better question than the ones asked before, that's a contribution, even if the answers elude us.

Which inexorably brings us to Matt Damon. In the movie version of the terrific book *The Martian*, Damon's character, Mark Watney, teaches astronauts-in-training what to do in case of impending doom. "At some point, everything's gonna go south on you and you're going to say, 'This is it. This is how I end,'" Watney says. You can either accept that as a failure—or you can get to work. "You do the math. You solve one problem. And you solve the next one. And then the next. If you solve enough problems, you get to come home."

If you solve enough problems, you get to land your rovers on Mars. If you solve enough problems, you get to build the Roman Empire. If you solve enough problems, you get to land on the Moon.

That's how you change the world. One problem at a time.

Changing the world one problem at a time requires delaying gratification. Most things in life are "first-order positive, second-order negative," as Shane Parrish writes on his website Farnam Street.[37] They give us pleasure in the short term but pain in the long. Spending money now instead of saving for retirement, using fossil fuels instead of renewable energy, guzzling sugar-laden beverages instead of water are all in that category.

When we're focused on first-order outcomes, we look for the instant success, the instant best seller, the instant fill-in-the-blank. We search for shortcuts, life hacks, and advice from self-proclaimed gurus. We "applaud the wrong things: the showy, dramatic record-setting sprint," Chris Hadfield writes, "rather than the years of dogged preparation or the unwavering grace displayed during a string of losses."[38] What's more, failure is expensive in the short term. When we're trying to maximize our profits and comfort *tomorrow*, we discount the value that failure brings in the long term. As a result, failure hits us hard. To boost our short-term pleasure, we avoid doing things that might fail.

Those who get ahead in life flip this perspective. "A real advantage is conferred on people who can do things that are first-order negative, second-order positive," Parrish writes.[39] These people delay gratification in a world that has become obsessed with it. They don't quit simply because their rocket blew up on the launch pad, they had a bad quarter, or their audition fell flat. They reorient their calibration for the long term, not for the short.

When it comes to creating long-lasting change, there are no hacks or silver bullets, as venture capitalist Ben Horowitz says. You'll need to use a lot of lead bullets instead.[40]

Inputs over Outputs

Think back on the failures you've had in your life. If you're like most people, you'll picture the bad outcomes—the business that never took off, the penalty kick you missed, or the job interview you bombed. Poker players, as Annie Duke explains in *Thinking in Bets*, refer to this tendency to "equate the quality of a decision with the quality of its outcome" as "resulting."[41] But as Duke argues, the quality of an input isn't the same as the quality of the output.

Focusing on outputs leads us astray because good decisions can lead to bad outcomes. In conditions of uncertainty, outcomes aren't completely within your control. An unforeseeable dust storm can cripple a perfectly designed Martian spacecraft. A bad wind can misdirect a perfectly shot soccer ball. A hostile judge or jury can derail a great case.

If we engage in resulting, we reward bad decisions that lead to good outcomes. Conversely, we change good decisions merely because they produced a bad outcome. We start shaking things up, reorganizing departments, or firing or demoting people. As one study shows, National Football League (NFL) coaches change their lineup after a one-point loss, but don't change it after a one-point win—even though these minor score differences are often poor indicators of player performance.[42]

Most of us act like American football coaches, treating success and failure as binary outcomes. But we don't live in a binary world. The line between success and failure is often razor thin. "Failure hovers uncomfortably close to greatness," wrote James Watson, the codiscoverer of DNA's double-helix structure.[43] The same decision that produced a failure in one scenario can lead to triumph in others.

The goal, then, is to focus on the variables you *can* control—the inputs—instead of the outputs. You should ask, "What went wrong with this failure?" and if the inputs need fixing, you should fix them. But this question isn't enough. You must also ask, "What went right with this failure?" You should retain the good-quality decisions, even if they produced a failure.

Consider Amazon's response to its Fire phone fiasco. Viewed through the standard metrics of output like profitability, the Fire was a colossal failure. But Amazon looked beyond the outcome. "When we try a new project, we look at the inputs," says Amazon's Andy Jassy.[44] "Did we hire a great team? Did the team have thoughtful ideas? Did they think the idea all the way through? Did they execute in a timely fashion? Was the quality high? Was the technology innovative?" Even if the project fails, you can take the inputs that worked and use them in future projects. "Not only did we take the learning from [the Fire] technology," Jassy explains, "but we also took all of the technology we built and applied it to a bunch of other services and capabilities."

Inputs aren't sexy. The word *input* might be better reserved for a boring database software. But an input-focused mind is the mark of anyone who has achieved anything extraordinary. The amateur focuses on getting hits and expects short-term results. The professional plays the long game and prioritizes inputs, perfecting them for years with no immediate payoff.

This is why the tennis player Maria Sharapova describes focus on outcomes as the worst mistake that beginning tennis players make.[45] Watch the ball as long as you can, Sharapova cautions, and zero in on the inputs. By taking the pressure off the outcome, you get better at your craft. Success becomes a consequence, not the goal.

This reorientation toward inputs has another upside. If you find yourself resenting the inputs, you might be chasing the wrong output. There's a question that frequently shows up in self-help books: What would you do if you knew that you could not fail? This isn't the right question to ask. Instead, do as Elizabeth Gilbert does, and flip the question on its head: "What would you do even if you knew that you might very well fail? What do you love doing so much that the words failure and success essentially become irrelevant?"[46] When we switch to an input-focused mindset, we condition ourselves to derive intrinsic value out of the activity. The input becomes its own reward.

With an input-focused mindset, you're free to change your destination. Goals can help you focus, but that focus can also turn into tunnel vision if you refuse to budge or pivot from your initial path.

For example, when Google Glass was roundly dismissed as a pointless product, X found a different path. Once the product hit the consumer market, the company realized that the Glass wasn't a consumer product at all. Instead, X learned from that failure and reinvented the Glass as a tool for businesses.[47] You can now find Google Glass on countless workers, including Boeing employees working on aircraft and doctors looking through a patient chart using a fancy attachment to their faces.[48]

Consider another example from the pharma industry. In 1989, Pfizer scientists developed a new drug called sildenafil citrate. Researchers hoped the drug would expand blood vessels to treat angina and high blood pressure associated with heart disease. By the early 1990s, the drug appeared to be ineffective for its intended purpose. But the participants in the trials reported an interesting side effect—erections. It wasn't long before researchers abandoned their initial hypothesis to pursue the astonishing alternative. And Viagra was born.[49]

Focusing on inputs has another upside. You avoid the wild swings of misery and euphoria that come with chasing outcomes. Instead, you become curious—no, fascinated—about tweaking and perfecting the inputs.

How Fascinating!

Mike Nichols was the prolific film director behind many classics, including *The Graduate*.[50] Although people tend to remember Nichols's hits, many of his films were flops. Some of these duds would appear from time to time—as flops do—on late-night television. Whenever Nichols came across one of his failures, he would park himself on his couch and watch the whole thing from start to finish.

As he sat and watched, what's important is what he wouldn't do. He wouldn't cringe. He wouldn't look away. He wouldn't blame the damn critics.

He'd simply watch and think, "That's so interesting, how that scene didn't work out." Not "I'm a loser." Not "This is awful." Not "What a complete embarrassment." Instead, with no judgment, he'd ponder, "Isn't it funny how sometimes things work and other times they don't?"

Nichols's approach reveals the secret to taking the sting out of failure. Curiosity takes a failure, turns the volume of drama all the way down, and makes failure interesting. It provides emotional distance, perspective, and an opportunity to view things through a different lens.

In their terrific book, *The Art of Possibility*, Rosamund Stone Zander and Benjamin Zander offer a practical method for putting this mindset to practice. Every time you make a mistake, every time you fail at something, you should throw your arms in the air and say, "How fascinating!"[51]

Fair warning: If you're anything like me, you'll grumble when you first do this. As you try to put your arms in the air, they'll go up ever so slowly—as if you're doing an imaginary bench press with really, really heavy weights. And the phrase "How fascinating!" will sound more petulant than joyous.

That's okay. Do it anyway. As you bask in the glory of your fascination, start asking some questions. What can I learn from this? What if this failure was actually good for me?

If you need inspiration, just picture Mike Nichols, sitting on his couch—not complaining about how the gods have turned on him by broadcasting his biggest failures on television for the world to see—but smiling, nodding, and knowing that watching this failure with curiosity means he'll do better the next time.

Flying Blind

Failure, as we've seen, is the portal to discovery, innovation, and long-term success. But most organizations suffer from collective amnesia over their failures. Mistakes remain concealed because employees are too afraid to share them. Most companies tell their employees, explicitly or implicitly, that if you succeed—according to short-term, quantifiable metrics like profits—you get a big pot of money, a better office, and a better title. If you fail, you get nothing. Or worse, you get shown the door.

This incentive scheme only exacerbates the deeply ingrained inertia against owning up to our failures. When we reward success and punish failure, employees will underreport failures, overreport successes, and reframe anything that falls in between in the best possible light. When we shoot the messengers, people stop delivering messages—particularly if they work for us. In one study, 42 percent of surveyed scientists across nine federal agencies (including NASA) feared retaliation for speaking out.[52] Of more than forty thousand employees surveyed at a tech company, 50 percent believed it wasn't safe to speak up at work.[53]

But failures transmit invaluable signals. Your goal should be to pick up these signals before your competitors do. But in most environments, these signals are elusive whispers that don't rise above the noise. If you can't hear them, if you suppress them, or if you shed them before they stick, you can't learn from them.

This is why airplanes carry flight recorders called black boxes. They record everything, including conversations in the cockpit and data from the airplane's electronic systems. The name *black box* is actually a misnomer because the box is bright orange to make it easier to find after a crash. The box is also fireproof, shockproof, and waterproof because the data it holds is crucial to uncovering why an accident happened.

We omit black boxes from our life to our detriment. Let's return for a moment to the crash of the 1999 Mars Polar Lander. Recall from earlier that the Lander crashed most likely because its engines shut down prematurely. But we don't know what happened for sure. Because money was tight, the lander lacked a way to communicate with mission control during its descent to the Martian surface. The team members had to cut corners, and this particular corner they cut deprived them—and all future rocket scientists—of the ability to extract critical lessons from this $120 million mishap.[54]

The omission resulted in part because the Mars Polar Lander was viewed myopically as a single project. If management had viewed the lander as part of a comprehensive whole—one probe among many interplanetary probes—then a communications device crucial for long-term learning should have been included.

To facilitate learning from failure, NASA catalogs mistakes in human spaceflight in a document called "Flight Rules."[55] The rules are a record of the past to guide the future. They're a body of knowledge culled from decades of missteps and miscalculations to ensure that the lessons endure. The document contains thousands of anomalies that have come up during manned spaceflight since the 1960s and the solutions to them. The book preserves this institutional knowledge for future generations, giving each failure shape and purpose as part of a larger story. It also obviates the need to reinvent the wheel and allows employees to focus on new problems. But as with any set of rules, these should be guardrails, not handcuffs. They should guide but not constrain. As we saw earlier, historical processes can rigidify into inflexible rules that impede first-principles thinking.

NASA's "Flight Rules" document works in part because the failures of others are the best catalyst for our own understanding. Our approach to failure is hypocritical. We explain away our own failures by blaming them on external factors. But when others stumble, we point to internal factors—they were careless, incompetent, or not paying enough attention. Our tendency to catalog the personal failings of others is why their mistakes can be a great source of learning. In one study, cardiac surgeons who observed their colleagues' blunders got significantly better at performing the procedure.[56] They homed in on the other surgeons' mistakes and learned not to repeat them.

Although companies pay lip service to tolerating and documenting failures, they often fail in practice. When I speak to corporate executives on failure, some of them argue that if failure is tolerated, then failures will multiply. Failure means fault, and fault needs to be assigned. If these executives don't discipline the responsible party, they assume they'll end up nurturing an anything-goes culture, where it becomes okay to fail.

These beliefs are out of step with a generation of research. As you'll see in the next section, you can create an environment of intelligent failures without complacency. You can allow people to take high-quality risks, but you can also set high standards. You don't have to tolerate sloppy failures—repeatedly making the same mistakes or failing because of a lack of care. You can reward intelligent failures, sanction poor performance, and accept that some errors are going to be inevitable when you're building things that may not work. People should be held accountable not for failing intelligently, but for failing to learn from it.

"There are two parts to failure," Pixar's former president Ed Catmull writes. "There is the event itself, with all its attendant disappointment, confusion, and shame, and then there is our reaction to it." We don't control the first part, but we do control the second. The goal, as Catmull puts it, should be "to uncouple fear and failure—to create an environment in which making mistakes doesn't strike terror into your employees' hearts."[57]

Rewarding intelligent failure sounds simple in theory, but it's difficult to implement in practice. A superficial commitment to "innovation" or "taking risks" won't create a culture of intelligent failures. In the next section, we'll explore how to create this ideal environment in the context of medicine, which provides a close analog to rocket science. The challenges on the operating room table aren't that different from those on the launch pad. The stakes are high. The pressure is on. The tiniest mistake can prove fatal. In this environment, creating a culture of intelligent failure is difficult, but as we'll see, it's not rocket surgery.

Psychological Safety

Medication errors in hospitals—where the wrong drug is given to the patient—are shockingly common. A 1995 study found 1.4 medication errors per patient per hospital stay. And of those errors, roughly 1 percent produced complications and harmed the patient.[58]

Amy Edmondson, a professor at Harvard Business School, wanted to explore the cause of these medication errors.[59] She asked herself, "Do better hospital teams make fewer medication mistakes?" To Edmondson, the answer seemed obvious. Better teams, with better-performing members and leaders, should make fewer errors.

But the results were the exact opposite. Better teams were making *more* mistakes, not less. What could explain this counterintuitive outcome?

Edmondson decided to dig deeper, sending a research assistant into the wild to observe the teams on the hospital floor. The assistant discovered that better teams weren't *making* more mistakes. Instead, they were simply *reporting* more mistakes. The teams that had a climate of openness—where the staff felt safe to discuss mistakes—performed better because employees were more willing to share failures and actively work to reduce them.

Edmondson refers to this climate as "psychological safety." I have to admit, when I first heard the term, I instinctively dismissed it as woo-woo. It conjured images of employees sitting around a conference table joining hands and sharing their feelings. But after studying the research, I backed down. The supporting evidence is rock solid. Psychological safety means, in Edmondson's words, "no one will be punished or humiliated for errors, questions, or requests for help, in the service of reaching ambitious performance goals."[60]

Research shows that psychological safety stimulates innovation.[61] When people feel free to speak up, ask provocative questions, and air half-formed thoughts, it becomes easier to challenge the status quo. Psychological safety also increases team learning.[62] In psychologically safe environments, employees challenge questionable calls by superiors instead of obediently complying with the commands.[63]

The best-performing hospital team in Edmondson's study was led by a hands-on, highly accessible nurse manager who actively facilitated an open environment. During interviews, the manager explained that a "certain level of error" is expected on her team and that a "nonpunitive environment" is essential to uncovering the error and addressing it. The nurses working in the unit confirmed the manager's statements. One noted that "people feel more willing to admit errors here, because [the manager] goes to bat for you." On this team, it was the nurses themselves who shouldered the responsibility for mistakes. As the manager explained, the "nurses tend to beat themselves up about errors; they are much tougher on themselves than I would ever be."[64]

The two worst-performing hospital teams had very different climates. In these teams, making a mistake meant getting punished. A nurse described one incident where she had inadvertently hurt a patient while drawing blood. The nurse manager put her "on trial," she explained, "it was degrading, like I was a two-year-old." Another nurse explained that "doctors condescend, and they bite your head off" if you make a mistake. One nurse said it was like "being called into the principal's office." As a result, if a medication error happened, nurses didn't advertise it, to save themselves from short-term embarrassment and anguish. By doing so, however, they discounted the long-term consequence of remaining silent—namely, injury or death of the patient because of a mistake.

This environment, in turn, led to a vicious cycle. The worst-performing teams—those that were in most need of improvement—were also the least likely to report errors. And if errors aren't reported, the team can't improve.

To encourage the reporting of failures, Google's moonshot factory, X, takes an unusual approach.[65] In most companies, it's a senior leader who decides to pull the plug on a faltering project. But employees at X are empowered to kill their own projects as soon as they realize, for one reason or another, that the project isn't viable.

Here's the interesting part: For this act of hara-kiri, the entire team receives a bonus. Recall from an earlier chapter that X led a project called Foghorn to convert seawater into fuel by sucking carbon dioxide out of the ocean water. Although the technology was promising, it wasn't economically viable, so the team decided to shut down its own project. "Thank you!" X's head, Astro Teller, announced at an all-hands meeting. "By ending their project, this team has done more to speed up innovation at X this month than any other team in this room."[66]

The notion of giving bonuses for failing might strike you as odd. It's one thing to tolerate failure, but something else to reward it. But there's genius in this incentive scheme. It's more expensive for nonviable projects to continue; they waste money and resources.[67] If a project has no future, shutting it down frees up precious resources for other moonshots that have better odds of landing. The resulting environments—where people are constantly generating intelligent failures—"remove the fear and make it safe for people to kill their project," X's Obi Felten explains.

Amazon follows a similar approach. If the quality of the inputs on a failed project was outstanding, then the team is rewarded through assignments in great new roles in the company—not punished. Otherwise, Amazon's Andy Jassy says, "you'll never get great people to take chances on new projects."[68]

This mindset translates to a six-word mantra: "Reward excellent failures, punish mediocre successes," as a seminar attendee once told author Tom Peters.[69] There must be a clear commitment to supporting intelligent failure and well-intentioned risk taking. People must know that intelligent failure is necessary for future success, that they won't be punished for it, and that their careers won't be ended for it. If the signals are mixed, employees will err on the side of caution and hide their mistakes instead of revealing them.

There's another component to psychological safety. If employees are to share their mistakes, the leaders must do the same.

Advertise Your Failures

It's not easy for smart, competitive people to own up to their blunders, particularly when no one else has noticed them. But astronauts are expected to advertise their own missteps and put them under a microscope for everyone to see.[70] Talking openly about screwups is mandatory because one astronaut's admission of a boneheaded move can save another's life.

Even where lives aren't at stake, advertising our failures can facilitate learning and develop psychological safety. This is why I started the "Famous Failures" podcast, where I interview the world's most interesting people about their failures and what they learned from them. As you might imagine, asking guests to appear on the show has made for some interesting conversations.

"Hey, Dan, I have a podcast where I interview failures. You'd be perfect for it."

Surprisingly though, most people I've approached have been eager to appear on the show because they know firsthand what many others fail to recognize: Anyone who has done anything meaningful has failed in some fashion. Having interviewed numerous titans on the podcast—including top entrepreneurs, Olympic medalists, and *New York Times* best-selling authors—I have found one common thread: Everyone—and I mean everyone—is a walking imperfection. Even genius isn't blunder-proof.

Einstein spoke openly about his biggest blunders. As astrophysicist Mario Livio writes, "More than 20 percent of Einstein's original papers contain mistakes of some sort."[71] Sara Blakely, the founder and CEO of Spanx, highlights her own oops moments at company-wide meetings.[72]

Catmull, the former president of Pixar, talks about the mistakes he has made at new-employee orientations: "We do not want people to assume that because we are successful, everything we do is right," he explains.[73] Economist Tyler Cowen wrote a detailed analysis of how, in the lead-up to the 2008 financial crisis, he "badly underestimated the chance that something systemic had gone wrong in the American economy." Cowen admitted his remorse: "I regret that I was wrong, and I regret that I was overconfident in my belief that I was right."[74]

If these individuals now appear more endearing to you, you're experiencing what researchers call the "beautiful mess effect."[75] Exposing your vulnerability can make you more desirable in the eyes of others. But there's one caveat. You must establish your competence before revealing your failures. Otherwise, you risk damaging your credibility and coming across as a mess—and not a beautiful one.[76]

Despite the beautiful-mess effect, most of us are terrible at owning up to our blunders. Our public image is synonymous with our self-worth. We puff ourselves up and create curated portrayals of our imperfect lives. We round off the edges, airbrush the negatives, and present a perfect image to the world devoid of any failures. Even when we talk about our failures, we do so in a flattering light.

I get it. It's painful to fail, and airing your failures can compound the pain. But the opposite approach—denial and avoidance—makes things worse. To learn and grow, we must acknowledge our failures without celebrating them.

This advice is particularly important for leaders. People pay close attention to the leader's behavior since they depend on the leader for recognition.[77] Studies also show that people look to the leader to initiate change.[78] If leaders fail to acknowledge their failures—if there's a perception that the leader can do no wrong—it's unrealistic to expect employees to take the risk of challenging the leader or revealing their own mistakes.

Consider a study of sixteen hospitals with top-tier cardiac surgery departments that implemented a new technology for surgery.[79] The technology upended how surgery was conducted. Each team had to unlearn ingrained habits and adopt different ones from scratch. The teams that learned quicker than the others shared three essential characteristics, one of which is particularly relevant here. They were led by surgeons who were more willing to acknowledge their own fallibility. For example, one surgeon repeatedly told his team, "I need to hear from you because I'm likely to miss things."[80] Another surgeon would say, "I screwed up. My judgment was bad in this case."

What made these messages effective was their repetition. Entrenched behaviors don't change with one impassioned speech. As team members heard these messages over and over again, they felt psychologically safe to speak up—even in an environment as hierarchical as an operating room. "There are no sacred cows," a member of one surgery team explained. "If somebody needs to be told something, then they are told—surgeon or orderly."[81]

Whether you're in the operating room, the boardroom, or the mission control room, the principle is the same. The road to success is filled with potholes. You're better off acknowledging them than pretending they don't exist.

How to Fail Gracefully

Not all failures are created equal. Some are more graceful than others. Rocket scientists use a constellation of tools to contain failures so the blunders don't create a cascade of damage. We covered some of these tools in earlier chapters. For example, rocket scientists conduct thought experiments where a failure produces no tangible damage. They build in redundancies so the mission doesn't fail even if a component fails. They use tests to lower the stakes because failures on the ground prevent far more disastrous ones in space.

Beyond rocket science, you can also use tests to fail more gracefully. Instead of rolling out an innovative policy across the entire company, you can use one division or a subset of your customers as a laboratory or an experiment. If one division breaks, the company still stands. If a subset of customers hates the policy, the damage is contained. For example, Starwood Hotels—which includes hotel brands like Westin and Sheraton—often used its W Hotels brand as an innovation lab, a testing ground for new ideas like signature scents and a living room experience in the hotel lobby. If the ideas worked in the smaller pilots in W Hotels, the company would roll them out to the other hotels in its portfolio.[82] If the ideas didn't work, the damage would be contained.

Testing has another upside. By definition, it allows you to practice failure in a relatively safe environment. Rocket scientists fail regularly, but for many of us—particularly in newer generations—failure can be an unfamiliar experience. As Jessica Bennett writes in the *New York Times*, "Faculty at Stanford and Harvard coined the term 'failure deprived' to describe what they were observing: the idea that, even as they were ever more outstanding on paper, students seemed unable to cope with simple struggles."[83]

Overcoming this fear requires exposure therapy. In other words, we must expose ourselves to failure regularly. Think of this as vaccination: Just as introducing weak antigens can stimulate "learning" in our immune system and prevent future infection, exposure to intelligent failures can allow us to recognize and learn from them. Each dose builds resilience and breeds familiarity. Each crisis becomes training for the next one.

This doesn't mean imposing catastrophic failures on ourselves. We don't have to be masochists. Rather, it means giving ourselves the breathing room to push boundaries, tackle thorny problems, and, yes, to fail. Let yourself fall on the grass. Give yourself permission to botch a song on the piano and write ghastly first drafts of book chapters (as I keep telling myself).

Parents can take a cue from Sara Blakely. She went from selling fax machines door-to-door to becoming the world's youngest self-made woman billionaire. She credits her success partly to a question that her father would ask her every week when she was growing up. "What have you failed at this week?" If Sara didn't have an answer, her father would be disappointed. To her father, failing to try was far more disappointing than failure itself.

....

WE OFTEN ASSUME that failure has an endpoint. We fail until we succeed and then stop failing to reap the benefits of our newly minted position in the pecking order. But failure isn't a bug to get out of our system until success arrives. Failure is the feature. If we don't develop a habit of failing regularly, we court catastrophe. As we'll see in the next chapter, where failure ends, complacency begins.

> Visit **ozanvarol.com/rocket** to find worksheets, challenges, and exercises to help you implement the strategies discussed in this chapter.

9

NOTHING FAILS
LIKE SUCCESS

How Success Produced the Biggest
Disasters in Rocket-Science History

If you can meet with Triumph and Disaster
And treat those two impostors just the same
...
Yours is the Earth and everything that's in it.
—RUDYARD KIPLING

"COME ON, ROGER. Come on in and watch."[1]
Roger Boisjoly wasn't in the mood for watching. Boisjoly (pronounced like the wine Beaujolais) was a mechanical engineer by training. He had spent a quarter century in the aerospace industry, first working on the Apollo lunar module and later joining a company called Morton Thiokol. At Thiokol, he served on the team that built the solid rocket boosters responsible for launching the space shuttle from the pad.

In July 1985, Boisjoly penned a memo that would prove to be prescient. The memo to his superiors warned about problems with the O-rings on the rocket boosters. O-rings are thin rubber bands that seal the joints of the boosters and prevent hot gases from leaking out of them. There were two O-rings on each joint—a primary and a secondary, for good measure—because the function they serve is critical. In several launches, engineers had discovered that both the primary and the secondary O-rings had been damaged. During a January 1985 mission, the primary O-ring failed, but the secondary O-ring saved the day after sustaining some damage itself. Boisjoly asked his superiors to take immediate action. He didn't mince his words: The result," he warned, "would be a catastrophe of the highest order, loss of human life."

On the evening of January 27, 1986—roughly six months after he penned the memo—Boisjoly sounded the warning bell one more time. Joined by other Morton Thiokol engineers, Boisjoly used a teleconference with NASA to push for a delay of a space shuttle launch that was scheduled for the next day. On that evening, the ordinarily balmy weather in Cape Canaveral, Florida—the launch site for the shuttle—had turned uncharacteristically cold, with temperatures dipping below freezing. Boisjoly and his engineer colleagues argued that the O-rings had to be flexible to perform their intended function and that they tended to turn brittle in cold weather. But the management at Thiokol and NASA overruled the engineers' recommendation.

"Come on, Roger. Come on in and watch."

The next morning, on January 28, his colleagues were nagging Boisjoly to join them in a room at Thiokol's management information center to watch the launch. Boisjoly finally relented. He swallowed his disapproval and reluctantly stepped into the center. At the time, a weather tower near the launch pad recorded the ambient temperature at 36°F. The temperature near the solid rocket booster joints—where the O-rings were located—was even colder, estimated at 28°F.

As the countdown neared zero, a surge of fear gripped Boisjoly. If the O-rings failed, they would fail at liftoff, he thought. This was the moment of truth. The solid rocket boosters ignited in a thunderous roar, and the shuttle began to inch up ever so slowly from the pad. When the shuttle cleared the launch tower, Boisjoly heaved a sigh of relief. "We just dodged a bullet," a colleague whispered to him.

As the space shuttle continued its upward climb, mission control beamed up a command to the crew to go to full power: "Go at throttle-up."

The crew replied: "Roger, go at throttle-up."

This was the last transmission received from the space shuttle *Challenger*. At about one minute into the flight, searing hot gases began to escape from the solid rocket boosters in a visible plume. Boisjoly's sigh of relief had been premature. The entire shuttle disintegrated in a cloud of smoke and molten debris, ultimately resulting in the deaths of all seven members of its crew. These images are seared into the minds of millions who had tuned in to watch the live event—in part because Christa McAuliffe, selected to be the first teacher in space, was on board the space shuttle.

A special commission was appointed by President Ronald Reagan—popularly known as the Rogers Commission after its chairman, William P. Rogers, former attorney general and secretary of state. The commission determined that the explosion resulted from a failure of the O-rings. At a commission hearing, Richard Feynman stunned television audiences by dropping an O-ring into ice water. The O-ring visibly lost its ability to seal in temperatures similar to those prevailing at the time of *Challenger*'s launch.

The recurring problems with the O-rings had been described in NASA documents as an "acceptable risk," the standard way of doing business. As one flight after another was completed despite dangerous levels of O-ring damage, NASA began to develop institutional tunnel vision. "Since the risk of O-ring erosion was accepted and indeed expected," explained NASA manager Lawrence Mulloy, "it was no longer considered an anomaly to be resolved before the next flight."[2]

The anomaly had become the norm. Feynman described NASA's decision-making process as "Russian roulette." Because no catastrophe had ensued after numerous shuttles flew with O-ring problems, NASA believed that "for the next flight we can lower our standards a little bit because we got away with it last time."[3]

It's easy to play Monday-morning quarterback and pretend it was obvious that the *Challenger* shouldn't have been launched. Hindsight tends to oversimplify and create the false impression that outcomes were inevitable. But even in hindsight, we can learn from these events, particularly because the *Challenger* accident and others I'll cover in this chapter replicate the same patterns of behavior that often arise in our personal and professional lives.

This chapter is about those lessons. I'll explain why it can be just as dangerous to celebrate success as it is to celebrate failure, and I'll reveal why a postmortem should follow both triumph and defeat. We'll explore why success is the wolf in sheep's clothing and how it conceals small failures that can snowball into the biggest disasters. You'll learn how a *Fortune* 500 company managed to stay ahead of the competition by reinventing itself twice and how you can disrupt yourself before others do it for you. You'll discover why the same type of flaw that produced the *Challenger* disaster also caused the 2008 collapse of the housing market, and you'll learn what German cab drivers and rocket scientists have in common. You'll leave the chapter with tactics for fending off complacency and learning from success.

Why Success Is a Lousy Teacher

Seventeen years after *Challenger*, it happened again.

Early Saturday morning on February 1, 2003, the space shuttle *Columbia* was on its way back home after spending sixteen days in space.[4] As the shuttle descended into the atmosphere at twenty-three times the speed of sound, the leading edges of its wings heated up to roughly 2,500°F because of expected atmospheric friction. But what wasn't expected was a series of erratic temperature readings. When mission control in Houston attempted to reach the astronauts, shuttle commander Rick Husband responded, "And, uh, Hou—" before cutting out. A second attempt by Husband to reach mission control was also cut short at "Roger." A minute later, all signals from *Columbia* were lost. Any hope that the signal loss simply resulted from malfunctioning sensors was dashed by live footage of the *Columbia* disintegrating on television. The flight director LeRoy Cain watched the footage in shock, unable to hold back the single tear sliding down his cheek. He collected himself and ordered "Lock the doors," starting the quarantine process that follows a disaster in space.

The space shuttle had blown up during reentry into the atmosphere, killing all seven astronauts on board and spreading debris over two thousand square miles. This time around, the culprit was a piece of foam insulation that was "about the size of a beer cooler."[5] During the launch, the foam had separated from the shuttle's external fuel tank and struck its left wing. The strike left a gaping hole in the thermal protection system responsible for protecting the shuttle from the searing heat during reentry.

A few days after the disaster, the space shuttle program manager downplayed the significance of the foam debris. Using language strikingly similar to his 1980s predecessors, he explained that foam debris had struck and damaged the shuttle in every mission. Over time, "foam shedding," as it was internally called at NASA, officially became an "accepted flight risk." James Hallock, an aviation safety expert and member of the Columbia Accident Investigation Board, explained that "not only was [foam shedding] expected, it eventually became accepted." It was formally described as an "in family" event, meaning "a reportable problem that was previously experienced, analyzed, and understood."[6]

Except that the problem wasn't understood. NASA had no idea why its shuttles were shedding foam, whether foam debris could compromise mission safety, or how it could be prevented.

Hallock took it upon himself to figure it out. He asked a simple question: How much force would it take to break the panels protecting the shuttle's wings from the heat of reentry? According to NASA specifications, the panels had to withstand a kinetic energy of 0.006 foot-pounds (a foot-pound is the energy required to raise one pound over one foot of distance). In a move reminiscent of Feynman's O-ring demonstration, Hallock conducted a simple experiment using a number 2 pencil and a small weight scale. He figured out that a pencil dropped from six inches would apply sufficient force to break the panels. To be sure, the panels were manufactured to be stronger than the specifications, but the low bar showed just how confident NASA was that nothing would strike the shuttle with sufficient force to compromise mission safety.

But the facts called this confidence into question. Roughly three months before the *Columbia* accident, the space shuttle *Atlantis* sustained a foam strike during its launch. The resulting damage was "the most severe of any mission yet flown."[7] Instead of suspending flights to investigate what happened, NASA marched ahead with *Columbia*'s launch.

The day after the launch, engineers conducting a routine review of launch videos noticed the foam strike. But the cameras in position to see the strike either didn't capture it or produced blurry images. Because of budget cuts, the camera lenses hadn't been properly maintained. Working with limited equipment, the engineers could tell that the "piece of foam was unusually large—larger than any they had seen."[8] But they could say no more.

When NASA structural engineer Rodney Rocha viewed the video and saw the size of the debris, he "gasped audibly."[9] He emailed his manager, Paul Shack, to determine whether the astronauts could inspect the impact area and perhaps repair it by performing a spacewalk. But he received no answer. Rocha later emailed Shack again to ask if NASA could "petition (beg) for outside agency assistance." The request was code for using the Pentagon's spy satellites to take images of the affected areas on the shuttle to survey the damage. In the email, Rocha outlined several options available to address the damage and safely land the shuttle. In other words, even a boss who tells her employees, "Don't just bring me problems; bring me the solutions," should have been pleased.

But Shack rebuffed Rocha's request. Shack later told Rocha that the management declined to pursue the matter. When Rocha pressed, Shack refused to relent: "I'm not going to be a Chicken Little on this," he said. Rocha and other concerned engineers were dismissively branded "foamologists" by NASA administrator Sean O'Keefe.

The senior management believed the foamologists were ringing the alarm bells over a routine event. Linda Ham, the chair of the Mission

Management Team, reminded her team that previous flights had been successfully completed despite foam strikes. "We haven't changed anything," she said, "we haven't experienced any 'safety of flight' damage in 112 flights." The shuttle, according to Ham, was "safe to fly with no added risk."[10]

This message was then beamed up to the *Columbia* crew. An email to the astronauts noted that the foam strike "was not even worth mentioning," but they should be informed in case they got a question from a reporter. The email concluded by reiterating that NASA had "seen this same phenomenon on several other flights and there is absolutely no concern for entry."[11]

Armed with this reassurance, the *Columbia* crew proceeded toward Earth. When the shuttle was just minutes away from its landing site, it broke apart after its battered thermal protection system allowed hot gases to penetrate the wing.

Science, as George Bernard Shaw writes, "becomes dangerous only when it imagines that it has reached its goal."[12] Before the *Challenger* accident, NASA had successfully launched shuttle missions despite the erosion of the O-rings. Before the *Columbia* accident, numerous shuttle launches had succeeded despite the shedding of foam. Each success reinforced a belief in the status quo. Each success fostered a damn-the-torpedoes attitude. With each success, what would otherwise be considered unacceptable levels of risk became the new norm.

Success is the wolf in sheep's clothing. It drives a wedge between appearance and reality. When we succeed, we believe everything went according to plan. We ignore the warning signs and the necessity for change. With each success, we grow more confident and up the ante.

But just because you're on a hot streak doesn't mean you'll beat the house.

As Bill Gates says, success is "a lousy teacher" because it "seduces smart people into thinking they can't lose."[13] Research supports this intuition.[14] In one representative study, financial analysts who made better-than-average predictions over four quarters grew overconfident and became less accurate with future predictions than their baseline.[15]

"Whom the Gods wish to destroy," wrote literary critic Cyril Connolly, "they first call promising."[16] The moment we think we've made it is the moment we stop learning and growing. When we're in the lead, we assume we know the answers, so we don't listen. When we think we're destined for greatness, we start blaming others if things don't go as planned. Success makes us think we have the Midas touch—that we can walk around turning everything into gold.

With the Apollo missions to the Moon, NASA had turned the impossible into the possible when the odds were heavily stacked against the agency. The successes blunted the most capable minds and boosted their egos. According to the Rogers Commission report, the improbable successes of the Apollo era produced a "We can do anything" attitude at NASA.[17]

But here's the thing: You can do some things wrong and still succeed. The technical term here is dumb luck. A spacecraft with a design flaw can safely land on Mars where the conditions don't trigger the flaw. A poorly shot soccer ball can end up in the goal if it ricochets off another player. A bad trial strategy can produce a win when the facts and the law are on your side.

But success has a way of concealing these blunders. When we're busy lighting cigars and popping champagne corks, we fail to account for the role that luck played in our triumph. Luck, as E. B. White put it, "is not something you can mention in the presence of self-made men."[18] Having worked hard to get to where we are, we resent the suggestion that anything other than elbow grease and talent produced the outcome. But when we fail to look in the mirror and recognize that we succeeded *despite* making a mistake and *despite* taking an unwise risk, we court catastrophe. The bad decisions and the dangers will continue into the future, and the success we once experienced will someday elude us.

This is why child prodigies unravel. This is why the housing market, believed to be the bedrock of the American economy, crumbled. This is why Kodak, Blockbuster, and Polaroid flamed out. In each case, the unsinkable sinks, the uncrashable crashes, and the indestructible self-destructs—because we assume their previous success secures their future.

Surviving your own success can be more difficult than surviving your own failure. We must treat success like a seemingly friendly group of Greeks bearing a big, beautiful gift called a Trojan horse. We must take measures to maintain humility before the Greeks arrive. We must treat our work—and ourselves—as permanent works in progress.

Permanent Works in Progress

In the early days of the space program, uncertainties loomed large. NASA was a newcomer, and its products—the Mercury, Gemini, and Apollo spacecraft—were decidedly works in progress. "We were so damned uncertain of what we were doing," explained NASA's chief engineer Milton Silveira. "We would ask for continual reviews, continual scrutiny by anybody we had respect for, to look at this thing and make sure we were doing it right."[19]

After the Apollo missions produced a string of resounding successes, the prevailing attitude at NASA began to change. The space agency, buoyed by the bureaucrats in Washington, began to view human spaceflight as routine. In January 1972, when the space shuttle program was announced, President Richard Nixon declared that the shuttle "will revolutionize transportation into near space, by *routinizing* it."[20] It was anticipated to be a reusable spacecraft that would fly frequently—as much as fifty times per year, according to initial estimates.[21] The shuttle would be a souped-up version of a Boeing 747 that "you could simply land and turn around and operate again."[22] Treating the shuttle like an airplane would have the additional benefit of attracting customers for payloads.

By November 1982, the shuttle "had proven sufficiently safe and error-free to become routine, reliable, and cost-effective," as two organizational researchers explain.[23] NASA was so confident in the safety of the space shuttle that, before the *Challenger* accident, the management saw no need to include an escape system for the crew.[24] And by the time of the *Challenger* mission, spaceflight was so routine that a civilian—an elementary school teacher—could ride shotgun to space.

As time wore on, NASA began to make compromises on safety and reliability. Its quality-assurance staff was cut by more than two-thirds, from roughly 1,700 in 1970 to 505 in 1986, the year that the *Challenger* was launched. Marshall Space Flight Center in Alabama—which is responsible for rocket propulsion—was the hardest hit, with a reduction from 615 to 88 staff members. The reductions meant "fewer safety inspections . . . less careful execution of procedures, less thorough investigation of anomalies, and less documentation of what happened."[25]

Routine also brought a standardized set of rules and procedures to NASA, with each flight becoming a straightforward application of those standards. Routine meant sticking to the previously scheduled programming and disregarding anomalies. NASA gradually morphed into a hierarchical organization where compliance with rules and procedures became more important than contribution.

The hierarchy also produced a disconnect between the engineers and the managers. The administrators at NASA abandoned the dirty-hands approach of the Apollo era. The managers were no longer intimately involved with the flight technology, and they eventually lost touch. The culture shifted from one focused on research and development to one that operated more like a business with production pressures.[26] The engineers were the ones with the dirty hands, and most of them still believed—despite what the bureaucrats were saying—that the space shuttle was a risky, experimental technology.[27] But the message didn't reach the top.

Let's return for a moment to the *Challenger* disaster. On the eve of the launch, Thiokol engineers argued that the *Challenger* shouldn't be launched unless the ambient temperature was above 53°F. But the shuttle program manager, Mulloy, balked. "What you are proposing to do," Mulloy said, "is to generate a new Launch Commit Criteria on the eve of launch, after we have successfully flown with the existing Launch Commit Criteria 24 previous times."[28] The assumption was that as long as the rules that produced previous successes were followed, nothing bad could result.

The moment we pretend an activity is routine is the moment we let our guard down and rest on our laurels. The remedy is to drop the word *routine* from our vocabulary and treat all our projects—particularly the successful ones—as permanent works in progress. NASA didn't lose a single crew member in space during the Apollo, Mercury, and Gemini missions, when human spaceflight was viewed as a risky work in progress. The only fatalities during those early years occurred during a launch rehearsal test on the ground, when the Apollo 1 spacecraft caught fire. It was only after human spaceflight was viewed as routine that we lost a NASA crew during flight. "We've grown used to the idea of space," President Reagan said after the *Challenger* disaster, "and perhaps we forget that we've only just begun."[29]

"Human beings," social psychologist Daniel Gilbert explains, "are works in progress that mistakenly think they're finished."[30] The five-time world track-and-field champion Maurice Greene didn't make that mistake and saw himself as a permanent work in progress. Even if you're a world champion, Greene would caution, you must train like you're number two.[31] When you're ranked second—or at least you pretend you are—you're less likely to grow complacent. You'll rehearse that speech until you know it cold, overprepare for that job interview, and work harder than your competitors.

This is why Bo Jackson, the only player to be named an all-star in both football and baseball, wouldn't get thrilled when he hit a home run or ran for a touchdown. He would say that he "hadn't done it *perfect*."[32] After his first hit in Major League Baseball, he defied tradition by refusing to keep the ball as a memento because, to Jackson, it was "just a ground ball up the middle."[33] Mia Hamm played soccer with the same mindset. "Many people say I'm the best women's soccer player in the world," Hamm once said. "I don't think so. And because of that, someday I just might be."[34] Charlie Munger, the business partner of Warren Buffett, uses the same approach as a rule of thumb in hiring decisions: "If you think your IQ is 160 but it's 150, you're a disaster. It's much better to have a 130 IQ and think it's 120."[35]

Research supports this approach. As Daniel Pink explains in *When*, "A team ahead at halftime—in any sport—is more likely than its opponent to win the game."[36] But there's an exception where motivation trumps mathematical reality. According to a study of over eighteen thousand professional basketball games, being slightly behind at halftime boosts a team's chances of winning.[37] The results also apply off the court and in the controlled environment of the laboratory. One study pitted participants against each other in a typing contest involving two periods separated by a short break.[38] During the break, participants were told they were either far behind (−50 points), slightly behind (−1 point), tied, or slightly ahead (+1 point). The participants who believed they were slightly behind exerted significantly more effort than all the other participants in the second period.

You can foster this never-complacent mindset by assuming you're trailing slightly behind and that the villain in your story—whether it's the Soviets for NASA, Hertz for Avis, or Nike for Adidas—is still in first place. When you ship a new product, you can explain how it can be improved in the next version. When you draft a memo or a book chapter, you can point out what's wrong with it.

The modern world doesn't call for finished products. It calls for works in progress, where perpetual improvement wins the game.

Success, Interrupted

Madonna is a master at reinventing herself. She has evolved with the times, collaborating with different producers and writers.[39] Her constant reinvention has been the hallmark of her superstardom for over three decades.

But major corporations aren't Madonnas. The wheels of corporate change are notoriously slow, particularly when it comes to fundamental transformations. But one major corporation managed to reinvent itself not just once—but twice—in record time.

Netflix started out by disrupting the traditional video rental model by shipping DVDs through the mail. But even as the company began to corner that market, its cofounder and CEO, Reed Hastings, remained vigilant.[40] As I discussed in an earlier chapter, we can reframe questions to generate better answers by focusing on strategy instead of tactics. Applying this principle, Netflix realized it wasn't in the DVD-delivery business. That was a tactic. Rather, it was in the movie-delivery business. That was its strategy. Delivering DVDs through the mail was simply one tactic among many others—including streaming media—in service of that strategy. "My greatest fear at Netflix," Hastings said, "has been that we wouldn't make the leap from success in DVDs to success in streaming."[41] Hastings saw the writing on the wall—DVDs would soon become obsolete—and tried to stay ahead of the melt on the ice-cream cone.

For Netflix, the leap to streaming arguably came too fast. In 2011, when the company announced plans to focus only on streaming and convert the DVD business to a separate, stand-alone company, its customers balked. But the mistake—if it was a mistake at all—was far better than the alternative of doing nothing. Hastings listened to his customers, picked up the pieces, and marched ahead with ramping up the company's streaming platform while retaining the DVD-by-mail business.

Netflix then made another leap into developing original content instead of paying the big studios in Hollywood for it. This leap turned out to be a huge success by every metric. Netflix had a disproportionate amount of hits compared with flops it ended up canceling. But to Hastings, this proportion was a bad omen. "Our hit ratio is too high right now," he said, "We should have a higher cancel rate overall."[42]

Hastings's desire for less success may strike you as irrational, but he's onto something. We often treat variances in our personal and professional success rates as errors. If given the option, we would prefer uninterrupted peak performance, rather than interruptions through the valleys of failure. But as business school professor Sim Sitkin explains, "regularity and uninterrupted success are a problem and a sign of weakness rather than an unequivocal sign of strength."[43]

Regular success, as the *Challenger* and *Columbia* disasters remind us, can portend long-term trouble. Research shows that success and complacency go hand in hand.[44] When we succeed, we stop pushing boundaries. Our comfort sets a ceiling, with our frontiers shrinking rather than extending. Corporate executives are rarely punished for deviating from a historically successful strategy. But the risk of punishment is far greater if an executive abandons a successful strategy to pursue one that ends up failing. As a result, instead of risking something new, we maintain the same "proven" formula that led to our success. This tactic works well—until it doesn't.

SpaceX's zero-for-three record for the Falcon 1 launches came close to killing the company, but those early failures served as sobering reality checks. They prevented the company from growing complacent. When those failures eventually gave way to a string of successes, SpaceX fell victim to its own hubris. In June 2015, a Falcon 9 rocket exploded on its way up to the International Space Station. Musk placed the blame squarely on the company's successful track record. "It's the first time we've had a failure in seven years," he said, "so, to some degree, the company as a whole got a little complacent."[45]

To prevent complacency, knock yourself off the pedestal once in a while. "You have to disrupt yourself," Steve Forbes says, "or others will do it for you."[46] If we don't experience variability in our track record—if we don't prevent our confidence from inflating after a string of random successes—then a catastrophic failure will do that for us. But catastrophic failures also tend to end your business or your career. "If you're not humble," said former world heavyweight champion Mike Tyson, "life will visit humbleness upon you."

One way to stay humble is to pay attention to near misses.

Near Misses

In aviation lingo, a near miss is an incident that could have been a hit. A near miss means you came close, but not close enough to cause a collision. It means you got lucky.

We tend to ignore near misses both in the air traffic control room and in the boardroom. Research shows that near misses masquerade as successes because they don't affect the ultimate outcome.[47] The airplane doesn't crash, the business doesn't tank, and the economy remains stable. *All's well that ends well*, and *no harm, no foul*, we tell ourselves and move on with our day.

It turns out that even if there's no harm, there can be plenty of foul. As we saw, NASA successfully launched numerous space shuttle missions despite the problems with the O-rings and despite the foam shedding. These earlier missions were *misses* because they didn't fail, but also *near* because luck intervened to save the day.[48]

Near misses lead people to take unwise risks. Rather than urgency, near misses create complacency. In studies, people who have information about near misses make riskier decisions than those with no information about them.[49] Even though the actual risk of failure remains the same after a near miss, our *perception* of the risk decreases.[50] At NASA, the management interpreted each near miss not as a potential problem, but as data that confirmed its belief that O-ring damage or foam shedding weren't risk factors and wouldn't compromise the mission. The managers had a perfect string of successes. The rocket scientists sounding the alarm were crying wolf.

The opposing data points didn't arrive until disaster struck. Only then did NASA gather the troops to conduct a postmortem and investigate the warnings that had been concealed by success. By then, it was too late.

A postmortem is a Latin phrase that literally means "after death." In a medical postmortem—also known as an autopsy—a dead body is examined to determine the cause of the death. Over the years, the term migrated from medicine to business. Companies now use a postmortem to determine why a failure happened and what can be done to prevent it in the future.

But there's a problem with this metaphor. A postmortem implies that there must be a dead project, a dead business, or a dead career before we're moved to action. The idea of death suggests that only catastrophic failures deserve a thorough investigation. But if we wait until disaster strikes to conduct a postmortem, the string of small failures and near misses—the chronic problems that build up slowly over time—go unnoticed.

Leading up to the *Columbia* and *Challenger* accidents, there wasn't one gross misjudgment, one major miscalculation, or one egregious breach of duty. Rather, "a series of seemingly harmless decisions were made that incrementally moved the space agency" to catastrophe, as sociologist Diane Vaughan writes.[51]

These were small steps, not giant leaps.

The story is a common one. Most corporations perish because they ignore the baby steps, the weak signals, the near misses that don't immediately affect outcomes. Merck, for example, ignored the early warning signs linking its painkiller Vioxx to cardiovascular disease.[52] Executives at Kodak ignored the signs that digital imaging could disrupt their business. Blockbuster paid little attention to the threat from Netflix's business model. There were signals that the subprime mortgage crisis was under way long before major financial institutions imploded in 2008 and generated one of the worst recessions in US history.

Consider also a study of over 4,600 orbital rocket launch attempts. According to the study, only total failures—where the rocket blew up—led to institutional learning and improved the likelihood of future success.[53] Partial or small failures—where the launch vehicle didn't blow up but failed to properly perform its function—had no similar effect. When small failures "are not widely identified, discussed, and analyzed, it is very difficult for larger failures to be prevented," as business school professors Amy Edmondson and Mark Cannon explain.[54]

Near misses are a rich source of data for a simple reason. They happen far more frequently than accidents. They're also significantly less costly. By examining near misses, you can gather crucial data without incurring the costs of failure.

Paying attention to near misses is particularly important in rocket science. Although rockets routinely exploded in the 1960s, we've gotten much better at getting them into space. The success rate of modern rockets is above 90 percent. Failure is the exception. But the stakes with each launch remain enormous. Hundreds of millions of dollars and, in human spaceflight, lives are at risk. What's more, failures in space often leave behind incomplete evidence. Much of the signal doesn't survive the noise, and it's hard to reproduce the failures on the ground. Where the learning opportunities from failure are few, it becomes all the more important to learn from success.

This leads to a paradox. We want our failures to be graceful so they don't wreak havoc on our lives. But graceful failures are also elusive failures, likely to escape notice unless we're paying close attention. The goal should be to spot these stealth signals before they snowball into something we can't control. This means that postmortems shouldn't be reserved for our worst days on the field. They should follow both failure and success.

The New England Patriots learned this lesson in the 2000 NFL draft.[55] The draft is an annual spectacle where football teams pick new players for the upcoming season. Each team gets to select one player in each of the seven rounds.

In the sixth round of the 2000 draft, the Patriots picked up a player who would go on to become one of the greatest quarterbacks of all time. Tom Brady would win six Super Bowls with the Patriots and pick up four Super Bowl Most Valuable Player awards—the most of any player in NFL history. Brady would be dubbed the "biggest steal" in the 2000 draft, and the Patriots leadership would be praised for its brilliant strategic maneuvering in scooping up a player of Brady's caliber at the tail end of the draft.[56]

That's one interpretation of the events.

Another interpretation is far less forgiving of the Patriots leadership. On an alternative reading, Brady was a near miss. The Patriots had their eye on him for a long time, but they waited until the end of the draft to pick him up (he was the 199th pick of 254 total players—almost as bad as getting picked last in gym class).[57] In an alternate universe, the same process could have generated a very different outcome. Another team could have drafted Brady before the Patriots did. Brady might not have realized his full potential if injuries hadn't crippled the Patriots' starting quarterback Drew Bledsoe, moving Brady into the starting lineup. In this alternate universe—which was inches from the actual one—the Patriots management would have been branded buffoons, not visionaries.

The next time you're tempted to start basking in the glory of your success while admiring the scoreboard, stop and pause for a moment. Ask yourself, What went *wrong* with this success? What role did luck, opportunity, and privilege play? What can I learn from it? If we don't ask these questions, luck will eventually run its course, and the near misses will catch up with us.

This set of questions, as you may have noticed, is no different from the ones we explored in the last chapter on failure. Asking the same questions and following the same process regardless of what happens is one way of taking the pressure off the outcome and reorienting our focus on what matters the most: the inputs.

Take your cue from X, Google's moonshot factory. Even when a technology succeeds, the engineers who worked on the products highlight the earlier prototypes that failed. For example, the team behind Project Wing—which developed self-flying delivery drones—had discarded hundreds of models before settling on the final design. At a company meeting, the team displayed its scrap pile—warts and all—for their colleagues to see. What appeared to the untrained eye as a simple brilliant design had emerged from a string of failures and near misses.[58]

The Patriots management knew it had gotten lucky with Brady. Instead of patting themselves on the back about their "biggest steal," the executives treated the Brady incident as a scouting failure and focused on fixing their mistakes.

A postmortem can be useful in uncovering and correcting mistakes. But it also has a drawback: When we conduct postmortems after a success, we already know the outcome. We tend to assume good outcomes resulted from good decisions and bad outcomes resulted from bad decisions. It's hard to find mistakes when we know we succeeded, and it's hard to avoid the blame game when we know we failed. Only when we blind ourselves to the high-beam lights of outcome can we more objectively assess our decision making.

Outcome Blind

A car racing team has its future on the line. It has been experiencing a series of inexplicable engine malfunctions. The engines have failed seven times in the last twenty-four races, causing serious damage to the car. The engine mechanic and the chief mechanic disagree about what's causing the problem.

The engine mechanic believes cold temperatures are to blame. When it's cold, the head and the block expand at different rates, he argues, damaging the gasket and causing the engine to fail. But the chief mechanic disagrees. He believes that temperature isn't the cause, because engine failures have occurred at all temperatures. The chief mechanic acknowledges that the drivers have their lives on the line during a race, but argues that in racing, "you are pushing the limits of what is known" and that if "you want to win, you have to take risks." He adds, "Nobody ever won a race sitting in the pits."

Today's race offers a lucrative sponsorship opportunity and substantial national television exposure. But the weather is unusually cold, and another engine failure would mean a reputational disaster.

What would you do? Would you race or sit this one out?

This scenario is from the Carter Racing case study that professors Jack Brittain and Sim Sitkin created for use as a learning tool in business school classes.[59] The students first decide individually what the racing team should do and then discuss the case study in class. Both before and after the class discussion, a vote is taken. Brittain and Sitkin report that roughly 90 percent of their students vote for proceeding with the race, citing some version of a "no guts, no glory" argument.

After the vote comes the punchline. The students are told, "You have just decided to launch the shuttle *Challenger*." The data on engine failures is similar to the data on the O-ring problems. There are other parallels as well—impending deadlines, budgetary pressures, as well as ambiguous and incomplete information.

When the punchline is delivered, most students express shock and, at times, anger. They feel tricked into making a decision that's obviously wrong and immoral. But the decision looks far less black-and-white when students are blinded to the outcome.

There are, of course, differences between the case study and the *Challenger* case. Although car engine failures could also compromise driver safety, the risk to human life isn't as acute in the case study as it is in a space shuttle launch.

But the moral still stands. It's easy for us to say we would have delayed the *Challenger* launch, drafted Brady in the first round, or seen the writing on the wall for Blockbuster. Concealing the outcome removes the distorting lenses of hindsight.

It's not easy to put blind analysis into practice outside a business school classroom. In the real world, outcomes aren't concealed. Once the cat is out of the bag, it's hard to put it back in. But there's a trick to putting blind analysis into practice without playing it stupid: the premortem.

The Premortem

Charlie Munger, the investor and partner of Warren Buffett, frequently quotes a "rustic" who says, "I wish I knew where I was going to die, and then I'd never go there."[60] This approach is called a premortem.[61] "There are two different occasions upon which we examine our own conduct," wrote Adam Smith, "and endeavour to view it in the light in which the impartial spectator would view it: first, when we are about to act; and secondly, after we have acted."[62] A postmortem covers Smith's second suggestion, and a premortem covers the first.

With a premortem, the investigation comes *before* we have acted, when the actual outcome isn't known—before we fire the rockets, close the sale, or complete the merger. In a premortem, we travel forward in time and set up a thought experiment where we assume the project failed. We then step back and ask, "What went wrong?" By vividly visualizing a doomsday scenario, we come up with potential problems and determine how to avoid them. According to research, premortems increase by 30 percent the ability of participants to correctly determine the reasons for a future outcome.[63]

If you're a business leader, a premortem might focus on a product you're currently designing. You would assume the product failed and then work backward to determine the potential reasons. Perhaps you didn't test the product properly or it wasn't the right fit for your market.

If you're a job candidate, a premortem might involve an interview. You would assume you didn't get the job and generate as many reasons as possible for the failure. Perhaps you were late for the interview. Perhaps a difficult question about why you left your previous job stumped you. You then figure out how to avoid those potential pitfalls.

Think of premortems as the opposite of backcasting, which we explored in the chapter on moonshot thinking. Backcasting works backward from a desired outcome. A premortem works backward from an undesired outcome. It forces you to think about what *could* go wrong before you act.

When you conduct a premortem and think through what can go wrong, you should assign probabilities to each potential problem.[64] If you quantify uncertainty ahead of time—there's a 50 percent chance that your new product might fail—you're more likely to recognize the role that luck played in any resulting success.

Quantifying uncertainty can also take the sting out of any failure that follows. If we're 100 percent confident that our new product will succeed, failure hits us hard. But if we recognize that there's only a 20 percent chance of success, failure won't necessarily mean the inputs were all bad. You could do everything right and still fail because luck and other factors intervene to tip the result.

Musk, for example, gave SpaceX less than a 10 percent chance of succeeding when he started the company.[65] His confidence was so low that he wouldn't let his friends invest. If he had given SpaceX, say, an 80 percent chance of success, it would have been more difficult to carry the momentum when the first three Falcon 1 launches failed. When the fate of SpaceX eventually turned around, this approach also made him realize the role of luck in the string of successes. "If things had just gone a little bit the other way," Musk says, "[SpaceX] would be dead."

The premortems we compile should be easily accessible. At X, these premortems "live on a site where anyone can post something that they're worried about going wrong in the future," explains Astro Teller.[66] The employees can post concerns about a specific project or the company as a whole. This approach builds institutional knowledge and guards against the sunk-cost bias. If we know there was uncertainty attached to a previous decision, it becomes easier to challenge it. "People are probably already saying these things in smaller groups," says Teller, "but they might not be saying it loudly, clearly, or often enough—often because these are things that might get you branded as a downer or disloyal."

NASA engineer Rodney Rocha had firsthand experience with being branded a downer or disloyal. His repeated requests for additional imagery to survey the damage caused by the foam strike on *Columbia* had been rebuffed by the management. While *Columbia* was still in orbit, he sat down at his computer and began writing an email to his superiors as a last-ditch effort.

"In my humble technical opinion," Rocha wrote, "this is the wrong (and bordering on irresponsible) answer. . . . I must emphasize (again) that severe enough damage . . . could present potentially grave hazards." He ended his email by typing, "Remember the NASA safety posters everywhere around stating, 'If it's not safe, say so'? Yes, it's that serious."

He saved the email as a draft. He never clicked send.

Rocha later told investigators that he didn't send the email because "he did not want to jump the chain of command" and that he felt he should "defer to management's judgment."[67] He had good reason to be worried. Roger Boisjoly, who had written the prescient memo predicting a disaster six months before *Challenger*, had paid a stiff price for blowing the whistle. After the *Challenger* disaster, Boisjoly testified before the Rogers Commission and turned over his memo, along with other internal documents, showing that his warnings fell on deaf ears at Thiokol. He was chastised by his colleagues and managers for airing the company's dirty laundry before the public.[68] "If you wreck this company," a former friend told him, "I'm going to put my kids on your doorstep."[69]

No one likes to be the skunk at a picnic, the lone holdout pounding her fists at the table. Skunks, like messengers, have a habit of getting shot. It's no wonder that groupthink pops up even in organizations whose lifeblood is creativity. Faced with potential backlash, we censor ourselves rather than go against the grain. We conform, rather than flout.

Success only exacerbates this tendency toward conformity. It drives overconfidence in the status quo, which in turn stifles dissent, precisely when dissent is most needed to prevent complacency. "Minority viewpoints are important," writes Berkeley psychologist Charlan Nemeth, a leading expert on groupthink, "not because they tend to prevail but because they stimulate divergent attention and thought."[70] Even when minority opinions are wrong, "they contribute to the detection of novel solutions and decisions that, on balance, are qualitatively better." In other words, dissenters force us to look beyond the dominant position, which tends to be the most obvious one.

Tragically, for the *Challenger* and *Columbia*, these dissenting voices were ignored.[71] The burden shifted to the engineers to prove their safety concerns with hard, quantifiable data. Instead of requiring proof that the spacecraft was safe to launch (*Challenger*) or safe to land (*Columbia*), engineers were required to prove that it wasn't safe. Roger Tetrault, a member of the Columbia Accident Investigation Board, explained the management's attitude toward the engineers in the following way: "Prove to me that it's wrong, and if you prove to me that there is something wrong, I'll go look at it."[72] But it didn't end there. The engineers were then denied the opportunity to make their case and prove their hypothesis. In the *Columbia* mission, for example, the managers rebuffed their requests for additional satellite imagery to survey the damage.

Premortems can be a powerful way of organically uncovering dissent. Because they assume a bad outcome—that the project failed—and ask people to generate reasons for the failure, they can provide psychological safety for expressing genuine criticism and relaying it upward.

The Cause Behind the Cause

There's a ritual that follows every catastrophe in space.

An accident board of experts is convened, witnesses are summoned, documents are gathered, flight data is parsed, wreckages are studied, and a somber report of findings and recommendations is drafted.

The tradition is in place not because history repeats itself. It rarely does. The chances are extremely low that faulty O-rings or foam shedding will cause another disaster in space.

No, the ritual takes place because history instructs. History informs. History, if you look carefully, can provide invaluable lessons. The ritual gives us a time to pause, to reassess and recalibrate, to learn and to change.

In the case of the *Challenger* disaster, two primary culprits emerged from the Rogers Commission report, one technical and the other human. The technical culprits were the O-rings that failed to properly seal. The human culprits were the NASA employees who made the egregious decision to fly the shuttle even though the O-rings could malfunction in cold temperatures.

In other words, the Rogers Commission focused on the first-order, or immediate, causes of the problem. The first-order causes are obvious. There's an intuitive appeal to attacking them. They're simpler to put on a PowerPoint or into a press release. They usually have a physical presence or a name. In the case of O-rings, the flaws can be fixed. In the case of NASA employees, they can be scapegoated, demoted, and fired.

But here's the problem: The causes of failure in a complex system—whether it's a rocket or a business—are usually multiple. Numerous factors, including technical, human, and environmental, might combine to produce the failure. Remedying only the first-order causes leaves the second- and third-order causes intact. These are the deeper causes lurking beneath the surface. They make the first-order causes happen and may lead to them again.

The deeper causes of the *Challenger* accident were hidden in NASA's dark underbelly, as unearthed by Diane Vaughan in her decisive account of the events. She explains that, contrary to the Rogers Commission's conclusions, the *Challenger* accident happened precisely because the managers did their jobs. They were following the rules—not violating them.

Vaughan uses the term "normalization of deviance" to describe this pathology. The prevailing culture at NASA had normalized flying with unacceptable risks. "The cultural understandings, rules, procedures, and norms that always had worked in the past did not work this time," Vaughan writes. "It was not amorally calculating managers violating rules that were responsible for the tragedy. It was conformity."[73] In other words, NASA didn't just have an O-ring problem. It also had a conformity problem.

The solutions to these deeper causes aren't sexy. A change in NASA's culture of conformity can't be televised. It doesn't make for good stump speeches. You can't dump conformity into ice water and watch it turn brittle during congressional hearings.

What's more, curing second- and third-order causes is much more difficult. It's easier to slap on a third O-ring on each joint (as NASA did after *Challenger*) than it is to cure the deeper cultural pathology prevalent in a massive bureaucracy.

But if we leave the deeper causes unaddressed, the cancer will keep coming back. This is why we heard, in astronaut Sally Ride's memorable words, the echoes of *Challenger* in the *Columbia* accident. As the only person to serve on the investigation boards for both accidents, Ride was uniquely qualified to draw this connection. The technical flaws in the two accidents were different, but the cultural flaws were similar. The deeper causes of the *Challenger* tragedy had remained unaddressed, even after the technical flaws were fixed and the key decision makers were replaced.

The remedy was a sleight of hand that provided the illusion of a cure. When we pretend that curing the first-order cause will also eliminate the second- and third-order causes, we end up masking them and exposing ourselves to future catastrophe. Treating the most obvious flaws gives us certainty and the satisfaction of doing something about the problem. But we're only playing a never-ending game of cosmic Whac-A-Mole. Once one problem is nailed down, another will pop up.

We do the same thing in our personal and professional lives. We take painkillers to cure our back pain. We believe we lost market share because of our competitors. We assume that foreign drug cartels are responsible for America's drug problem and that eradicating the Islamic State group will solve terrorism.

In each case, we confuse a symptom with a cause and leave the deeper causes intact. Painkillers won't cure our back pain; the source remains. You're losing market share not because of your competitors but because of your own business policies. Eliminating cartels won't solve the demand side of the drug problem, and eradicating terrorists won't prevent new ones from cropping up.

Killing the Bad One often gives rise to the Worse One. In attacking the most visible causes, we unleash a Darwinian process of creating a more insidious pest. When the pest returns, we apply the same pesticide, up the dosage, and express shock when nothing changes.

A quote by George Santayana seems to appear in every museum depicting historical horrors: "Those who cannot remember the past are condemned to repeat it."[74] But remembering isn't enough. History is an exercise in self-deception if we get the wrong messages from it. Only through the hard work of looking beyond the first-order causes— particularly when we're afraid of what we might see—do we begin to learn from history.

Treating only the first-order causes has yet another downside. As we'll see in the next section, it can exacerbate, rather than solve, the problem.

The Unsafety of Safety

I'm not a morning person. To me, sunrises feel as energizing as a root canal. To prepare myself for what felt like a recurring battle each morning, I would set my alarm clock thirty minutes fast.

You know the rest of the story. Kid, meet snooze button. In economics lingo, I would *consume* those thirty minutes instead of *saving* them by repeatedly hitting snooze.

There is a phenomenon that explains my love-hate relationship with the snooze button. The same phenomenon shows why head and neck injuries increased in American football after players started wearing hard-shelled helmets to better protect themselves. It explains why installing antilock brakes—a now-ancient technology introduced in cars in the 1980s to avoid skidding—didn't decrease the number of accidents. It also explains why marking crosswalks doesn't necessarily make crossing the street any safer. In some cases, it leads to more fatalities and injuries.

The psychologist Gerald Wilde calls this phenomenon risk homeostasis.[75] The phrase is fancy, but the idea is simple. Measures intended to decrease risk sometimes backfire. Humans compensate for the reduced risk in one area by increasing risk in another.

Consider, for example, a three-year study conducted in Munich.[76] One portion of a taxicab fleet was equipped with an antilock brake system (ABS). The remainder of the cabs had traditional, non-ABS brakes. The cars were identical in all other respects. They drove at the same time of day, the same days of the week, and in the same weather conditions. The drivers also knew whether their car was equipped with ABS.

The study found no tangible difference in accident rates between the ABS-equipped cars and the remainder. But one difference was statistically significant: driving behavior. The drivers of the ABS-equipped cars became far more reckless. They tailgated more often. Their turns were sharper. They drove faster. They switched lanes dangerously. They were involved in more near misses. Paradoxically, a measure introduced to boost safety promoted unsafe driving behavior.[77]

Safety measures also backfired in the *Challenger* mission. The managers believed that O-rings had a sufficient safety margin "to enable them to tolerate three times the worst erosion observed up to that time."[78] What's more, there was a fail-safe in place. Even if the primary O-ring failed, the officials assumed the secondary O-ring would seal and pick up the slack.[79] The existence of these safety measures boosted a sense of invincibility and led to catastrophe when both the primary and the secondary O-rings failed during launch. These rocket scientists were like German cabbies in ABS-equipped cars, driving fast and loose.

In each case, the "safe" felt safer than it actually was. The corresponding behavior change eliminated any benefit from the safety measure. Sometimes, the pendulum swung in the other direction: The activity became less safe than it was before the safety measure was put in place.

This paradox doesn't mean that we stop fastening our seat belts, buy ancient cars that don't come with ABS, or take up jaywalking. Instead, pretend the crosswalk isn't marked, and walk accordingly. Assume the secondary O-ring or the ABS brakes won't prevent the accident. Keep your head out of the tackle, even if you're wearing a helmet. Act as if you didn't receive an extension on that project deadline.

The safety net *may* be there to catch you if you fall, but you're better off pretending it doesn't exist.

Visit **ozanvarol.com/rocket** to find worksheets, challenges, and exercises to help you implement the strategies discussed in this chapter.

EPILOGUE

THE NEW WORLD

Up, up the long, delirious, burning blue
I've topped the wind-swept heights with easy grace.
Where never lark, or even eagle flew—
And, while with silent, lifting mind I've trod
The high untrespassed sanctity of space,
Put out my hand, and touched the face of God.
—JOHN MAGEE

I N *THE SIMPSONS* episode "Deep Space Homer," Homer Simpson is engaged in his favorite pastime—channel surfing—when he comes across a space shuttle launch. As two monotonous commentators explain how the crew will explore the effects of weightlessness on tiny screws, Homer loses all interest. He tries to change the channel, but the batteries drop out of the remote control. A frantic Bart begins screaming, "Not another boring space launch. Change the channel. Change the channel!" The episode then cuts to the NASA headquarters, where a concerned rocket scientist explains to an administrator that they have run into a serious problem with the mission: The TV ratings for the launch are the lowest ever.

In 1994, when this episode aired, the heyday of human space exploration had become a distant memory. It took a dizzying six and a half decades from the Wright brothers' first powered flight in 1903 to humankind's first footsteps on the Moon in 1969. Yet, in the next five decades, we stopped looking up. We planted a flag and returned home, preferring to send humans into low Earth orbit for repeated trips to the International Space Station. For many, after watching Apollo astronauts brave the roughly 239,000-mile voyage to the Moon, seeing astronauts fly 240 miles up to the station was as thrilling as "watching Columbus sail to Ibiza."[1]

Politicians used spaceflight for political ends, effectively hanging a guillotine over NASA's head. Ambitious missions were announced in John F. Kennedy fashion by one administration only to be canceled by the next. Funding waxed and waned in response to the prevailing political winds. As a result, NASA lacked a clear vision. In 2012, shortly before his death, Neil Armstrong reportedly invoked the baseball legend Yogi Berra to describe the agency's predicament: "If you don't know where you are going, you might not get there."[2]

We didn't know where we would go after NASA retired the space shuttle in 2011—our only means of reaching the International Space Station—with no replacement in place. After the remaining space shuttles rolled off the launch pads and into museums, American astronauts had to ride shotgun to the station on Russian rockets. The tickets cost $81 million per passenger—nearly $20 million more than the launch of an entire SpaceX Falcon 9 rocket.[3] In an ironic twist, the space agency, established to beat the Russians, became dependent on them. After the United States imposed sanctions on Russia for annexing Crimea in 2014, Dmitry Rogozin—the deputy prime minister responsible for the Russian space program—threatened to retaliate by suggesting that "the US bring their astronauts to the [Station] using a trampoline."[4]

NASA's facilities became the physical embodiment of this state of affairs. In May 2014, when NASA tweeted photos of astronauts training in the Neutral Buoyancy Lab—the giant indoor pool that simulates a microgravity environment—the images were notable for what they did *not* show. Omitted from the photos was a large cordoned-off part of the pool that was leased to oil-services companies to conduct survival training for rig workers.[5] Also missing from the photo was the aftermath of a corporate party that had taken place the night before, with the pool serving as a backdrop. Launchpad 39A at Kennedy Space Center—one of the two historic pads from where the Apollo missions took off for the Moon—was abandoned and put up for lease.[6] What would have been a first-ever women-only spacewalk scheduled for March 2019 was canceled for lack of properly fitting spacesuits for the two women selected for the spacewalk.[7]

In the movie *Apollo 13*, a congressman asks Jim Lovell, the commander of the mission, "why we're continuing to fund this program now that we've beaten the Russians to the moon." Lovell, played by Tom Hanks, replies, "Imagine if Christopher Columbus came back from the New World, and no one returned in his footsteps."

NASA was the reason I—like many others—fell in love with space exploration. For decades, the acronym NASA represented the gold standard for thinking like a rocket scientist. Yet, after it blazed a trail to the New World, NASA largely handed off the human-spaceflight baton to others. In 2004, while the space shuttle was still grounded in the wake of the *Columbia* disaster, Burt Rutan's *SpaceShipOne* became the first privately funded vehicle to reach space.[8] Then, after the space shuttle was officially retired, NASA awarded SpaceX and Boeing contracts to build rockets to take American astronauts to the International Space Station. In a symbolic turn of events, SpaceX moved into Launchpad 39A and began launching its rockets from there.[9] Blue Origin is building its own road to space with its rockets—called *New Shepard* and *New Glenn*—named after the first American space pioneers: the Mercury astronauts Alan Shepard and John Glenn. The company is also building a lunar lander, named *Blue Moon*, capable of delivering cargo to the Moon. Although NASA is also working on a vehicle to launch humans beyond Earth orbit—called Space Launch System, or SLS—the effort is massively underfunded and behind schedule. As a result, its critics have dubbed the SLS the "rocket to nowhere."[10]

In one scene in the movie *The Wizard of Oz*, Dorothy steps out of her house to see the world in its full-color glory for the first time after spending her life living in monochrome. Once she sees the vivid colors, she can't unsee them. For her, there's no going back to black-and-white.

But the world doesn't work this way. Our default mode is regress—not progress. When left to their own devices, space agencies decline. Writers wither. Actors flare out. Internet millionaires collapse under the weight of their egos. Young and scrappy companies turn into the same acronym-driven, bloated bureaucracies they were seeking to displace. We return to black-and-white.

The journey cannot end once the mission is accomplished. That's when the real work begins. When success brings complacency—when we tell ourselves that now that we've discovered the New World, there's no reason to return—we become a shadow of our former selves.

In every annual letter to Amazon shareholders, Jeff Bezos includes the same cryptic line: "It remains Day 1." After repeating this mantra for a few decades, Bezos was asked what Day 2 would look like. He replied, "Day 2 is stasis. Followed by irrelevance. Followed by excruciating, painful decline. Followed by death. And *that* is why it is *always* Day 1."[11]

The rocket-science mindset requires remaining in Day 1 and repeatedly introducing color into the monochromatic world. We must keep devising thought experiments, taking moonshots, proving ourselves wrong, dancing with uncertainty, reframing problems, testing as we fly, and returning to first principles.

We must keep walking the untrodden paths, sailing the wild seas, and flying the savage skies. "However sweet these laid-up stores, however convenient this dwelling, we cannot remain here," Walt Whitman wrote. "However shelter'd this port and however calm these waters, we must not anchor here."[12]

In the end, there's no hidden playbook. No secret sauce. The power is there for the taking. Once you learn how to think like a rocket scientist—and nurture that thinking in the long term—you can turn the unimaginable into the imaginable, mold science fiction into fact, and stretch out your hands to touch the face of God.

The powerful play goes on, to quote Whitman again, and you may contribute a verse.

A new verse.

Even a whole new story.

Your story.

What will it say?

WHAT'S NEXT?

NOW THAT YOU'VE learned how to think like a rocket scientist, it's time to act like one and put these principles into action.

Head over to ozanvarol.com/rocket to find the following:

- A summary of key points from each chapter
- Worksheets, challenges, and exercises to help you implement the strategies discussed in the book
- A sign-up form for my weekly newsletter, where I share additional tips and resources that reinforce the principles in the book (readers call it "the one email I look forward to each week")
- My personal email address so you can share comments or just say hello!

I travel the globe frequently to give keynote talks to organizations across numerous industries. If you're interested in inviting me to talk to your group, please visit ozanvarol.com/speaking.

I look forward to hearing from you.

ACKNOWLEDGMENTS

THIS BOOK WOULD not exist without Steve Squyres, my former supervisor and the principal investigator of the Mars Exploration Rovers mission.

I don't know what compelled Steve to offer a job to a skinny kid with a funny name from a country halfway across the globe, but I'm so grateful that he did. It was my privilege to work with him alongside the rest of the team at Cornell, to whom I will remain forever grateful.

There have been a handful of mentors in my life who have dramatically altered its course for the better. Adam Grant is one of them. In October 2017, when I was navigating the unfamiliar territory of nonacademic book publishing, Adam referred me to his literary agent, Richard Pine. Within forty-eight hours of Adam's referral, Richard and I agreed to work together, which triggered the series of events that led to the publication of this book. Adam is a true giver and embodies what he preaches in his first book, *Give and Take: Why Helping Others Drives Our Success*. He has had an indelible impact on my life, and I'm fortunate to be one of his mentees and friends.

Adam introduced Richard as "the world's best literary agent." He wasn't kidding. Richard championed this book and helped me shape the loose ideas floating around my mind into a compelling package. I sleep much better knowing that Richard has my back. A heartfelt thank you to the rest of the all-stars at InkWell—including Alexis Hurley and Eliza Rothstein.

I'm deeply grateful to many mentors and colleagues for their sage advice about publishing—including Susan Cain, Tim Ferriss, Seth Godin, Julian Guthrie, Ryan Holiday, Isaac Lidsky, Barbara Oakley, Gretchen Rubin, and Shane Snow. Special thanks to Daniel Pink, who gave me an invaluable Book Publishing 101 lesson over coffee in Portland and came up with the subtitle for this book.

Thank you to my terrific editor at PublicAffairs, Benjamin Adams, who was a major creative force behind this book. It was a pleasure working with the entire PublicAffairs team, including Melissa Veronesi, Lindsay Fradkoff, Miguel Cervantes, and Pete Garceau.

Any author would be lucky to work with a copy editor as skilled as Patricia Boyd of Steel Pencil Editorial. She improved almost every sentence in this book with her amazing red pen.

Thank you to the wonderful people who let me interview them—including Mark Adler, Peter Attia, Natalya Bailey, Obi Felten, Tim Ferriss, Patrick Lieneweg, Jaime Waydo, Julie Zhuo, along with others who chose to remain anonymous. I'm grateful to Dina Kaplan and Baya Voce for making introductions. Thank you also to Libby Leahy, Head of Communications at X, and James Gleeson, Communications Director at SpaceX, for helping me verify relevant facts.

I'm deeply grateful to Nicholas Lauren and Kristen Stone, who went above and beyond with their comments on the first draft of this manuscript. Kristen sat at our dining room table and read her favorite quotes from the book out loud, giving me a small glimpse of what it must feel like to have the audience sing along with your band.

I'm fortunate to work with a wonderful team of people. My research assistant Kelly Muldavin provided editorial and research guidance. Brendan Seibel, Sandra Cousino Tuttle, and Debbie Androlia diligently checked countless facts and sources (any remaining mistakes are entirely mine). Michael Roderick of Small Pond Enterprises provided invaluable marketing and business counsel and saved me from numerous missteps. The incredible Brandi Bernoskie and her talented team at Alchemy+Aim designed beautiful web pages for this book and my other endeavors.

I'm grateful to the listeners of my podcast, Famous Failures, and the readers of my newsletter, the *Weekly Contrarian* (you can join at weekly contrarian.com). Special thanks to the members of my Inner Circle, a group of my most engaged readers, for letting me try out new ideas on them.

Our Boston terrier, Einstein, lives up to his name with his curiosity and wits. Thank you for filling our house with chew toys and our hearts with joy.

My parents, Yurdanur and Tacettin, gave me my first astronomy lessons and encouraged me to pursue an education in the United States even though it meant their only child would live thousands of miles away. *Hayatım boyunca beni desteklediğiniz için çok teşekkür ederim.*

Finally, Kathy, my wife, my best friend, my first reader—my first everything. Kurt Vonnegut once said, "Write to please just one person." For me, that person is Kathy. Thank you for talking with me about every idea in this book, reading the early drafts, laughing at the jokes, and being by my side during the peaks and troughs. Without you, my small steps could never become giant leaps.

....

THE FOLLOWING IS a list of early supporters who committed to pre-ordering and helping promote the book. I'm deeply grateful to them.

Cagatay Akkoyun, Janette Atkins, Sean Kevin Barry, Charles A. Bly, Jessica Bond, William Brent, Catherine Cheng MD, Timothy J. Chips, Doug Claffey, Luci Englert McKean, Rishi Ganti, Christina Guthier, Hannah and Thea, Rebecca Hartenbaum Davis, Raymond Hornung, James D. Kirk, Ramesh Kumar, Mark Kwesi Appoh, Nicholas Lauren, Jennifer J. LeBlanc, Jennifer LeTourneau, Susan Litwiller Ed.S., Maggie the Chihuahua, Kathleen Marie, Tony Martignetti, Sharon Mork, Terry Oehler, Julian Olin and Gaby Porras, Orthogon Partners, Tim Oslovich, Joe Pasquale, Cory Peticolas, Jack W. Richards, Noel Rudie, Hans-Dieter Schulte, Javier Segovia P., Renu Sharma, Brian Thompson, Jonathan "Bong" Valdez, and Vista Capital Partners.

NOTES

Introduction

1. NASA, "First American Spacewalk," National Aeronautics and Space Administration (hereafter cited as NASA), June 3, 2008, www.nasa.gov/multimedia/imagegallery/image_feature_1098.html.

2. Bob Granath, "Gemini's First Docking Turns to Wild Ride in Orbit," NASA, March 3, 2016, www.nasa.gov/feature/geminis-first-docking-turns-to-wild-ride-in-orbit.

3. Rod Pyle, "Fifty Years of Moon Dust: Surveyor 1 Was a Pathfinder for Apollo," NASA Jet Propulsion Laboratory, California Institute of Technology, June 2, 2016, www.jpl.nasa.gov/news/news.php?feature=6523; David Kushner, "One Giant Screwup for Mankind," *Wired*, January 1, 2007, www.wired.com/2007/01/nasa.

4. Stanley A. McChrystal et al., *Team of Teams: New Rules of Engagement for a Complex World* (New York: Portfolio, 2015), 146.

5. Robert Kurson, *Rocket Men: The Daring Odyssey of Apollo 8 and the Astronauts Who Made Man's First Journey to the Moon* (New York: Random House, 2018), 48, 51.

6. Kurson, *Rocket Men*, 48, 51.

7. John F. Kennedy, address at Rice University, Houston, September 12, 1962.

8. Andrew Chaikin, "Is SpaceX Changing the Rocket Equation?," *Air and Space Magazine*, January 2012, www.airspacemag.com/space/is-spacex-changing-the-rocket-equation-132285884/?page=2.

9. Kim Dismukes, curator, "The Amazing Space Shuttle," NASA, January 20, 2010, https://spaceflight.nasa.gov/shuttle/upgrades/upgrades5.html.

10. Lyrics to Elton John's "Rocket Man (I Think It's Going to Be a Long, Long Time)" can be found on the Genius website, https://genius.com/Elton-john-rocket-man-i-think-its-going-to-be-a-long-long-time-lyrics.

11. Stuart Firestein, *Ignorance: How it Drives Science* (New York: Oxford University Press, 2012), 83.

12. Carl Sagan, *Broca's Brain: Reflections on the Romance of Science* (New York: Random House, 1979), 15.

13. Nash Jenkins, "After One Brief Season, *Cosmos* Makes Its Final Voyage," *Time*, June 9, 2014, https://time.com/2846928/cosmos-season-finale.

14. Ben Zimmer, "Quants," *New York Times Magazine*, May 13, 2010, www.nytimes.com/2010/05/16/magazine/16FOB-OnLanguage-t.html.

15. Marshall Fisher, Ananth Raman, and Anna Sheen McClelland, "Are You Ready?," *Harvard Business Review*, August 2000, https://hbr.org/2000/07/are-you-ready.

16. Bill Nye, *Everything All at Once: How to Unleash Your Inner Nerd, Tap into Radical Curiosity, and Solve Any Problem* (Emmaus, PA: Rodale Books, 2017), 319.

17. Carl Sagan and Ann Druyan, *Pale Blue Dot: A Vision of the Human Future in Space* (New York: Random House, 1994), 6.

Chapter 1: Flying in the Face of Uncertainty

1. Lunar and Planetary Institute, "What is ALH 84001?," Universities Space Research Association, 2019, www.lpi.usra.edu/lpi/meteorites/The_Meteorite.shtml.

2. Vincent Kiernan, "The Mars Meteorite: A Case Study in Controls on Dissemination of Science News," *Public Understanding of Science* 9, no. 1 (2000): 15–41.

3. "Ancient Meteorite May Point to Life on Mars," CNN, August 7, 1996, www.cnn.com/TECH/9608/06/mars.life.

4. "Pres. Clinton's Remarks on the Possible Discovery of Life on Mars (1996)," video, YouTube, uploaded July 2, 2015, www.youtube.com/watch?v=pHhZQWAtWyQ.

5. David S. McKay et al., "Search for Past Life on Mars: Possible Relic Biogenic Activity in Martian Meteorite ALH84001," *Science*, August 16, 1996, https://science.sciencemag.org/content/273/5277/924.

6. Michael Schirber, "The Continuing Controversy of the Mars Meteorite," *Astrobiology Magazine*, October 21, 2010, www.astrobio.net/mars/the-continuing-controversy-of-the-mars-meteorite; Jasen Daley, "Scientists Strengthen Their Case That a Martian Meteorite Contains Signs of Life," *Popular Science*, June 25, 2010, www.popsci.com/science/article/2010-06/life-mars-reborn.

7. Peter Ray Allison, "Will We Ever . . . Speak Faster Than Light Speed?," BBC, March 19, 2015, www.bbc.com/future/story/20150318-will-we-ever-speak-across-galaxies.

8. Jet Propulsion Laboratory, "Past Missions: Ranger 1–9," NASA, www2.jpl.nasa.gov/missions/past/ranger.html.

9. R. Cargill Hall, "The Ranger Legacy," in *Lunar Impact: A History of Project Ranger*, NASA History Series (Washington, DC: NASA, 1977; website updated 2006), https://history.nasa.gov/SP-4210/pages/Ch_19.htm.

10. Steve W. Squyres, *Roving Mars: Spirit, Opportunity, and the Exploration of the Red Planet* (New York: Hyperion, 2005), 239–243, 289.

11. Yuval Noah Harari, *21 Lessons for the 21st Century* (New York: Spiegel & Grau, 2018).

12. The section on Fermat's Last Theorem draws on the following sources: Stuart Firestein, *Ignorance: How It Drives Science* (New York: Oxford University Press, 2012); Simon Singh, *Fermat's Last Theorem: The Story of a Riddle That Confounded the World's Greatest Minds for 358 Years* (London: Fourth Estate, 1997); NOVA, "Solving Fermat: Andrew Wiles," interview with Andrew Wiles, PBS, October 31, 2000, www.pbs.org/wgbh/nova/proof/wiles.html; Gina Kolata, "At Last, Shout of 'Eureka!' in Age-Old Math Mystery," *New York Times*, June 24, 1993, www.nytimes.com/1993/06/24/us/at-last-shout-of-eureka-in-age-old-math-mystery.html; Gina Kolata, "A Year Later, Snag Persists in Math Proof," *New York Times*, June 28, 1994, www.nytimes.com/1994/06/28/science/a-year-later-snag-persists-in-math-proof.html; John J. Watkins, *Number Theory: A Historical Approach* (Princeton, NJ: Princeton University Press, 2014), 95 (2013); Bill Chappell, "Professor Who Solved Fermat's Last Theorem Wins Math's Abel Prize, NPR, March 17, 2016, www.npr.org/sections/thetwo-way/2016/03/17/470786922/professor-who-solved-fermat-s-last-theorem-wins-math-s-abel-prize.

13. Kolata, "At Last, Shout of 'Eureka!'"

14. "Origins of General Relativity Theory," *Nature*, July 1, 1933, www.nature.com/articles/132021d0.pdf.

15. David J. Gross, "The Discovery of Asymptotic Freedom and the Emergence of QCD," Nobel Lecture, December 8, 2004, www.nobelprize.org/uploads/2018/06/gross-lecture.pdf.

16. US Department of Defense, "DoD News Briefing: Secretary Rumsfeld and Gen. Myers," news transcript, February 12, 2002, https://archive.defense.gov/Transcripts/Transcript.aspx?TranscriptID=2636; CNN, "Rumsfeld / Knowns," video of Rumsfeld statement on February 12, 2002, YouTube, uploaded March 31, 2016, www.youtube.com/watch?v=REWeBzGuzCc.

17. Donald Rumsfeld, author's note in *Known and Unknown: A Memoir* (New York: Sentinel, 2010), available at papers.rumsfeld.com/about/page/authors-note.

18. Errol Morris, "The Anosognosic's Dilemma: Something's Wrong but You'll Never Know What It Is (Part 1)," *New York Times*, June 20, 2010, https://opinionator.blogs.nytimes.com/2010/06/20/the-anosognosics-dilemma-1.

19. Daniel J. Boorstin, *The Discoverers: A History of Man's Search to Know His World and Himself* (New York: Random House, 1983).

20. Mario Livio, *Brilliant Blunders: From Darwin to Einstein—Colossal Mistakes by Great Scientists That Changed Our Understanding of Life and the Universe* (New York: Simon & Schuster, 2013), 140.

21. Derek Thompson, *Hit Makers: The Science of Popularity in an Age of Distraction* (New York: Penguin, 2017).

22. Sir A. S. Eddington, *The Nature of the Physical World* (Cambridge: Cambridge University Press, 1948), available at http://henry.pha.jhu.edu/Eddington .2008.pdf.

23. Brian Clegg, *Gravitational Waves: How Einstein's Spacetime Ripples Reveal the Secrets of the Universe* (Icon Books, 2018), 150–152; Nola Taylor Redd, "What Is Dark Energy?," *Space.com*, May 1, 2013, www.space.com/20929-dark-energy .html.

24. NASA, "Dark Energy, Dark Matter," NASA Science, updated July 21, 2019, https://science.nasa.gov/astrophysics/focus-areas/what-is-dark-energy.

25. James Clerk Maxwell, *The Scientific Letters and Papers of James Clerk Maxwell*, vol. 3, *1874–1879* (New York: Cambridge University Press 2002), 485.

26. George Bernard Shaw, toast to Albert Einstein, October 28, 1930.

27. Albert Einstein, *Ideas and Opinions: Based on Mein Weltbild* (New York: Crown, 1954), 11.

28. Alan Lightman, *A Sense of the Mysterious: Science and the Human Spirit* (New York: Pantheon Books, 2005).

29. The discussion on Steve Squyres is based on the following sources: Squyres, *Roving Mars*; University of California Television, "Roving Mars with Steve Squyres: Conversations with History," video, YouTube, uploaded August 18, 2011, www .youtube.com/watch?v=NI6KEzsb26U&feature=youtu.be; Terri Cook, "Down to Earth With: Planetary Scientist Steven Squyres," *Earth Magazine*, June 28, 2016, www.earthmagazine.org/article/down-earth-planetary-scientist-steven-squyres.

30. Steven Spielberg, in *Spielberg*, documentary film (HBO, 2017).

31. Richard Branson, "Two-Way Door Decisions," Virgin, February 26, 2018, www.virgin.com/richard-branson/two-way-door-decisions.

32. Ernie Tretkoff, "Einstein's Quest for a Unified Theory," *American Physical Society News*, December 2005, www.aps.org/publications/apsnews/200512 /history.cfm; Walter Isaacson, *Einstein: His Life and Universe* (New York: Simon & Schuster, 2007).

33. Jim Baggott, "What Einstein Meant by 'God Does Not Play Dice,'" *Aeon*, November 21, 2018, https://aeon.co/ideas/what-einstein-meant-by-god-does -not-play-dice.

34. Tretkoff, "Einstein's Quest."

35. Kent A. Peacock, "Happiest Thoughts: Great Thought Experiments in Modern Physics," in *The Routledge Companion to Thought Experiments*, ed. Michael T. Stuart, Yiftach Fehige, and James Robert Brown, Routledge Philosophy

Companions (London and New York: Routledge/Taylor & Francis Group, 2018).

36. A. B. Arons and M. B. Peppard, "Einstein's Proposal of the Photon Concept: A Translation of the Annalen der Physik Paper of 1905," *American Journal of Physics* 33 (May 1965): 367, www.informationphilosopher.com/solutions/scientists/einstein/AJP_1905_photon.pdf.

37. Charles Darwin, *On the Origin of Species by Means of Natural Selection* (New York: D. Appleton and Company, 1861), 14.

38. John R. Gribbin, *The Scientists: A History of Science Told Through the Lives of Its Greatest Inventors* (New York: Random House, 2004).

39. Richard P. Feynman, *The Pleasure of Finding Things Out: The Best Short Works of Richard P. Feynman* (New York: Basic Books, 2005) (emphasis in original).

40. The discussion of William Herschel's discovery of Uranus is based on the following sources: Emily Winterburn, "Philomaths, Herschel, and the Myth of the Self-Taught Man," *Notes and Records of the Royal Society of London* 68, no. 3 (September 20, 2014): 207–225, www.ncbi.nlm.nih.gov/pmc/articles/PMC4123665; Martin Griffiths, "Music(ian) of the Spheres: William Herschel and the Astronomical Revolution," *LabLit*, October 18, 2009, www.lablit.com/article/550; Ken Croswell, *Planet Quest: The Epic Discovery of Alien Solar Systems* (New York: Free Press, 1997), 34–41; Clifford J. Cunningham, *The Scientific Legacy of William Herschel* (New York: Springer Science+Business Media, 2017), 13–17; William Sheehan and Christopher J. Conselice, *Galactic Encounters: Our Majestic and Evolving Star-System, From the Big Bang to Time's End* (New York: Springer, 2014), 30–32.

41. William Herschel, *The Scientific Papers of Sir William Herschel*, vol. 1 (London: Royal Society and the Royal Astronomical Society, 1912), xxix–xxx.

42. Ethan Siegel, "When Did Isaac Newton Finally Fail?," *Forbes*, May 20, 2016, www.forbes.com/sites/startswithabang/2016/05/20/when-did-isaac-newton-finally-fail/#8c0137648e7e; Michael W. Begun, "Einstein's Masterpiece," *New Atlantis*, fall 2015, www.thenewatlantis.com/publications/einsteins-masterpiece.

43. Ethan Siegel, "Happy Birthday to Urbain Le Verrier, Who Discovered Neptune with Math Alone," *Forbes*, March 11, 2019, www.forbes.com/sites/startswithabang/2019/03/11/happy-birthday-to-urbain-le-verrier-who-discovered-neptune-with-math-alone/#6674bcd7586d.

44. Clegg, *Gravitational Waves*, 29.

45. Clegg, *Gravitational Waves*, 29.

46. Isaacson, *Einstein: His Life and Universe*.

47. T. C. Chamberlin, "The Method of Multiple Working Hypotheses," *Science*, May 1965, http://arti.vub.ac.be/cursus/2005-2006/mwo/chamberlin1890science.pdf.

48. Isaac Asimov, "The Relativity of Wrong," *Skeptical Inquirer* 14 (fall 1989):35–44.

49. Thomas S. Kuhn, *Structure of Scientific Revolutions* (Chicago: University of Chicago Press, 1962), xxvi.

50. Howard Wainer and Shaun Lysen, "That's Funny . . . A Window on Data Can Be a Window on Discovery," *American Scientist*, July 2009, www.american scientist.org/article/thats-funny.

51. For the discovery of quantum mechanics, see John D. Norton, "Origins of Quantum Theory," online chapter in Einstein for Everyone course, University of Pittsburgh, fall 2018, www.pitt.edu/~jdnorton/teaching/HPS_0410/chapters /quantum_theory_origins. For X-rays, see Alan Chodos, ed., "November 8, 1895: Roentgen's Discovery of X-Rays," This Month in Physics History series, *American Physical Society News* 10, no. 10 (November 2001), www.aps.org/publications/aps news/200111/history.cfm. For DNA, see Leslie A. Pray, "Discovery of DNA Structure and Function: Watson and Crick," *Nature Education* 1, no. 1 (2008): 100, www.nature.com/scitable/topicpage/discovery-of-dna-structure-and-function -watson-397. For oxygen, see Julia Davis, "Discovering Oxygen, a Brief History," *Mental Floss*, August 1, 2012, http://mentalfloss.com/article/31358/discovering -oxygen-brief-history. For penicillin, see Theodore C. Eickhoff, "Penicillin: An Accidental Discovery Changed the Course of Medicine," *Endocrine Today*, August 2008, www.healio.com/endocrinology/news/print/endocrine-today/%7B15 afd2a1-2084-4ca6-a4e6-7185f5c4cfb0%7D/penicillin-an-accidental-discovery -changed-the-course-of-medicine.

52. Andrew Robinson, *Einstein: A Hundred Years of Relativity* (Princeton, NJ: Princeton University Press, 2015), 75.

53. The section on the discovery of Pluto is based on the following sources: Croswell, *Planet Quest*; Michael E. Brown, *How I Killed Pluto and Why It Had It Coming* (New York: Spiegel & Grau, 2010); Kansas Historical Society, "Clyde Tombaugh," modified January 2016, www.kshs.org/kansapedia/clyde-tombaugh /12222; Alok Jha, "More Bad News for Downgraded Pluto," *Guardian*, June 14, 2007, www.theguardian.com/science/2007/jun/15/spaceexploration.starsgalaxies andplanets; David A. Weintraub, *Is Pluto a Planet? A Historical Journey through the Solar System* (Princeton, N.J.: Princeton University Press, 2014), 144.

54. NASA, "Eris," NASA Science, https://solarsystem.nasa.gov/planets/dwarf -planets/eris/in-depth.

55. Paul Rincon, "Pluto Vote 'Hijacked' in Revolt," BBC, August 25, 2006, http://news.bbc.co.uk/2/hi/science/nature/5283956.stm.

56. Robert Roy Britt, "Pluto Demoted: No Longer a Planet in Highly Controversial Definition," *Space.com*, August 24, 2006, www.space.com/2791-pluto -demoted-longer-planet-highly-controversial-definition.html.

57. A. Pawlowski, "What's a Planet? Debate over Pluto Rages On," CNN, August 24, 2009, www.cnn.com/2009/TECH/space/08/24/pluto.dwarf.planet/index.html.

58. American Dialect Society, "'Plutoed' Voted 2006 Word of the Year," January 5, 2007, www.americandialect.org/plutoed_voted_2006_word_of_the_year.

59. "My Very Educated Readers, Please Write Us a New Planet Mnemonic," *New York Times*, January 20, 2015, www.nytimes.com/2015/01/20/science/a-new-planet-mnemonic-pluto-dwarf-planets.html.

60. ABC7, "Pluto Is a Planet Again—At Least in Illinois," ABC7 Eyewitness News, March 6, 2009, https://abc7chicago.com/archive/6695131.

61. Laurence A. Marschall and Stephen P. Maran, *Pluto Confidential: An Insider Account of the Ongoing Battles Over the Status of Pluto* (Dallas: Benbella Books, 2009), 4.

62. Smithsonian National Air and Space Museum, "Exploring the Planets," https://airandspace.si.edu/exhibitions/exploring-the-planets/online/discovery/greeks.cfm.

63. Ralph Waldo Emerson, *The Essential Writings of Ralph Waldo Emerson* (New York: Modern Library, 2000), 261.

64. *In the Shadow of the Moon*, directed by Dave Sington (Velocity/Think Film, 2008), DVD.

65. Virginia P. Dawson and Mark D. Bowles, eds., *Realizing the Dream of Flight* (Washington, DC: NASA History Division, 2005), 237.

66. Mary Roach, *Packing for Mars: The Curious Science of Life in the Void* (New York: W.W. Norton, 2010).

67. Chris Hadfield, *An Astronaut's Guide to Life on Earth: What Going to Space Taught Me About Ingenuity, Determination, and Being Prepared for Anything* (New York: Little, Brown and Company, 2013).

68. Caroline Webb, *How to Have a Good Day: Harness the Power of Behavioral Science to Transform Your Working Life* (New York: Crown Business, 2016), 258.

69. Anne Fernald and Daniela K. O'Neill, "Peekaboo Across Cultures: How Mothers and Infants Play with Voices, Faces, and Expectation," in *Parent-Child Play: Descriptions and Implications*, ed. Kevin MacDonald (Albany: State University of New York Press, 1993).

70. Fernald and O'Neill, "Peekaboo Across Cultures."

71. W. Gerrod Parrott and Henry Gleitman, "Infants' Expectations in Play: The Joy of Peek-a-boo," *Cognition and Emotion* 3, no. 4 (January 7, 2008), www.tandfonline.com/doi/abs/10.1080/02699938908412710.

72. James Luceno and Matthew Stover, *Labyrinth of Evil, Revenge of the Sith, and Dark Lord: The Rise of Darth Vader*, The Dark Lord Trilogy: Star Wars Legends (New York: DelRey Books, 2011), 562–563.

73. For more on redundancy, see Shane Parrish, "An Introduction to the Mental Model of Redundancy (with Examples)," Farnam Street (blog), July 2011, https://fs.blog/2011/07/mental-model-redundancy.

74. SpaceX, "Falcon 9," www.spacex.com/falcon9; Andrew Chaikin, "Is SpaceX Changing the Rocket Equation?," *Air and Space Magazine*, January 2012, www .airspacemag.com/space/is-spacex-changing-the-rocket-equation-132285884/?no -ist=&page=2.

75. Tim Fernholz, *Rocket Billionaires: Elon Musk, Jeff Bezos, and the New Space Race* (Boston: Houghton Mifflin Harcourt, 2018); Dan Leone, "SpaceX Discovers Cause of October Falcon 9 Engine Failure," *SpaceNews*, December 12, 2012, https://spacenews.com/32775spacex-discovers-cause-of-october-falcon-9-engine -failure.

76. Hadfield, *Astronaut's Guide*.

77. James E. Tomayko, "Computers in the Space Shuttle Avionics System," in *Computers in Spaceflight: The NASA Experience* (Washington, DC: NASA, March 3, 1988), https://history.nasa.gov/computers/Ch4-4.html; United Space Alliance, LLC, "Shuttle Crew Operations Manual," December 15, 2008, www .nasa.gov/centers/johnson/pdf/390651main_shuttle_crew_operations_manual.pdf.

78. Scott Sagan, "The Problem of Redundancy Problem: Why More Nuclear Security Forces May Produce Less Nuclear Security," *Risk Analysis* 24, no. 4 (2004): 938, http://citeseerx.ist.psu.edu/viewdoc/download?doi=10.1.1.128.3515& rep=rep1&type=pdf.

79. NASA, "NASA Will Send Two Robotic Geologists to Roam on Mars," NASA press release, June 4, 2003, https://mars.nasa.gov/mer/newsroom/press releases/20030604a.html.

80. University of California Television, "Roving Mars with Steve Squyres."

81. A. J. S. Rayl, "Mars Exploration Rovers Update: Spirit Mission Declared Over, Opportunity Roves Closer to Endeavour," *Planetary Society*, May 31, 2011, www.planetary.org/explore/space-topics/space-missions/mer-updates/2011/05 -31-mer-update.htm.

82. Stephen Clark, "Scientists Resume Use of Curiosity Rover's Drill and Internal Lab Instruments," *Spaceflight Now*, June 5, 2018, https://spaceflightnow .com/2018/06/05/scientists-resume-use-of-curiosity-rovers-drill-and-internal -lab-instruments.

83. Neel V. Patel, "The Greatest Space Hack Ever: How Duct Tape and Tube Socks Saved Three Astronauts," *Popular Science*, October 8, 2014, www.popsci .com/article/technology/greatest-space-hack-ever.

Chapter 2: Reasoning from First Principles

1. The opening section on Elon Musk draws on the following sources: Tim Fernholz, *Rocket Billionaires: Elon Musk, Jeff Bezos, and the New Space Race*

(Boston: Houghton Mifflin Harcourt, 2018); Ashlee Vance, *Elon Musk: Tesla, SpaceX, and the Quest for a Fantastic Future* (New York: Ecco, 2015); Chris Anderson, "Elon Musk's Mission to Mars," *Wired*, October 21, 2012, www.wired.com/2012/10/ff-elon-musk-qa/all; Tim Fernholz, "What It Took for Elon Musk's SpaceX to Disrupt Boeing, Leapfrog NASA, and Become a Serious Space Company," *Quartz*, October 21, 2014, https://qz.com/281619/what-it-took-for-elon-musks-spacex-to-disrupt-boeing-leapfrog-nasa-and-become-a-serious-space-company; Tom Junod, "Elon Musk: Triumph of His Will," *Esquire*, November 15, 2012, www.esquire.com/news-politics/a16681/elon-musk-interview-1212; Jennifer Reingold, "Hondas in Space," *Fast Company*, February 1, 2005, www.fastcompany.com/52065/hondas-space; "Elon Musk Answers Your Questions! SXSW, March 11, 2018," video, YouTube, uploaded March 11, 2018, www.youtube.com/watch?v=OoQARBYbkck; Tom Huddleston Jr., "Elon Musk: Starting SpaceX and Tesla Were 'the Dumbest Things to Do,'" CNBC, March 23, 2018, www.cnbc.com/2018/03/23/elon-musk-spacex-and-tesla-were-two-of-the-dumbest-business-ideas.html.

2. Reingold, "Hondas in Space."

3. Adam Morgan and Mark Barden, *Beautiful Constraint: How to Transform Your Limitations into Advantages, and Why It's Everyone's Business* (Hoboken, NJ: Wiley, 2015), 36–37.

4. Darya L. Zabelina and Michael D. Robinson, "Child's Play: Facilitating the Originality of Creative Output by a Priming Manipulation," *Psychology of Aesthetics, Creativity, and the Arts* 4, no. 1 (2010): 57–65, www.psychologytoday.com/files/attachments/34246/zabelina-robinson-2010a.pdf.

5. Robert Louis Stevenson, *Robert Louis Stevenson: His Best Pacific Writings* (Honolulu: Bess Press, 2003), 150.

6. Yves Morieux, "Smart Rules: Six Ways to Get People to Solve Problems Without You," *Harvard Business Review*, September 2011, https://hbr.org/2011/09/smart-rules-six-ways-to-get-people-to-solve-problems-without-you.

7. Jeff Bezos, Letter to Amazon Shareholders, 2016, Ex-99.1, SEC.gov, www.sec.gov/Archives/edgar/data/1018724/000119312517120198/d373368dex991.htm.

8. Andrew Wiles, quoted in Ben Orlin, "The State of Being Stuck," Math with Bad Drawings (blog), September 20, 2017, https://mathwithbaddrawings.com/2017/09/20/the-state-of-being-stuck.

9. Micah Edelson et al., "Following the Crowd: Brain Substrates of Long-Term Memory Conformity," *Science*, July 2011, www.ncbi.nlm.nih.gov/pmc/articles/PMC3284232; Tali Sharot, *The Influential Mind: What the Brain Reveals About Our Power to Change Others* (New York: Henry Holt and Co., 2017), 162–163.

10. Gregory S. Berns et al., "Neurobiological Correlates of Social Conformity and Independence During Mental Rotation," *Biological Psychiatry* 58, no. 3 (2005): 245–253.

11. Astro Teller, "The Secret to Moonshots? Killing Our Projects," *Wired*, February 16, 2016, www.wired.com/2016/02/the-secret-to-moonshots-killing -our-projects.

12. Terence Irwin, *Aristotle's First Principles* (New York: Oxford University Press, 1989), 3.

13. *New World Encyclopedia*, s.v. "methodic doubt," updated September 19, 2018, www.newworldencyclopedia.org/entry/Methodic_doubt.

14. The discussion on how SpaceX used first-principles thinking draws on the following sources: Junod, "Elon Musk: Triumph of His Will"; Anderson, "Elon Musk's Mission to Mars"; Andrew Chaikin, "Is SpaceX Changing the Rocket Equation?," *Air and Space Magazine*, January 2012, www.airspacemag.com/space /is-spacex-changing-the-rocket-equation-132285884/?no-ist=&page=2; Johnson Space Center Oral History Project, https://historycollection.jsc.nasa.gov/JSC HistoryPortal/history/oral_histories/oral_histories.htm; Reingold, "Hondas in Space"; Fernholz, "Disrupt Boeing, Leapfrog NASA"; Fernholz, *Rocket Billionaires*.

15. Tom Junod, "Elon Musk: Triumph of His Will," *Esquire*, November 15, 2012, www.esquire.com/news-politics/a16681/elon-musk-interview-1212.

16. Johnson Space Center Oral History Project, "Michael J. Horkachuck," interviewed by Rebecca Wright, NASA, November 6, 2012, https://historycollection .jsc.nasa.gov/JSCHistoryPortal/history/oral_histories/C3PO/HorkachuckMJ /HorkachuckMJ_1-16-13.pdf.

17. The section on reusability in rocket science draws on the following sources: Fernholz, *Rocket Billionaires*; Tim Sharp, "Space Shuttle: The First Reusable Spacecraft," *Space.com*, December 11, 2017, www.space.com/16726-space-shuttle .html; Chaikin, "Changing the Rocket Equation?"; "Elon Musk Answers Your Questions!"; Loren Grush, "Watch SpaceX Relaunch Its Falcon 9 Rocket in World First," *Verge*, March 31, 2017, www.theverge.com/2017/3/31/15135304 /spacex-launch-video-used-falcon-9-rocket-watch; SpaceX, "X Marks the Spot: Falcon 9 Attempts Ocean Platform Landing," December 16, 2014, www.spacex .com/news/2014/12/16/x-marks-spot-falcon-9-attempts-ocean-platform-landing; Loren Grush, "SpaceX Successfully Landed Its Falcon 9 Rocket After Launching It to Space," Verge, December 21, 2015, www.theverge.com/2015/12/21/10640306 /spacex-elon-musk-rocket-landing-success.

18. Fernholz, *Rocket Billionaires*, 24.

19. SpaceX, "X Marks the Spot."

20. Elizabeth Gilbert, *Eat Pray Love: One Woman's Search for Everything* (New York: Viking, 2006).

21. Alan Alda, "62nd Commencement Address," Connecticut College, New London, June 1, 1980, https://digitalcommons.conncoll.edu/commence/7.

22. Nassim Nicholas Taleb, *Antifragile: Things That Gain from Disorder* (New York: Random House, 2012), 308.

23. The section on Steve Martin is based on Steve Martin, *Born Standing Up: A Comic's Life* (New York: Scribner, 2007), 111–113.

24. Dawna Markova, *I Will Not Die an Unlived Life: Reclaiming Purpose and Passion* (Berkeley, CA: Conari Press, 2000).

25. Anaïs Nin, *The Diary of Anaïs Nin*, ed. Gunther Stuhlmann, vol. 4, *1944–1947* (New York: Swallow Press, 1971).

26. Shellie Karabell, "Steve Jobs: The Incredible Lightness of Beginning Again," *Forbes*, December 10, 2014, www.forbes.com/sites/shelliekarabell/2014/12/10 /steve-jobs-the-incredible-lightness-of-beginning-again/#35ddf596294a.

27. Henry Miller, *Henry Miller on Writing* (New York: New Directions, 1964), 20.

28. The discussion on Alinea draws on the following sources: Sarah Freeman, "Alinea 2.0: Reinventing One of the World's Best Restaurants: Why Grant Achatz and Nick Kokonas Hit the Reset Button," *Eater.com*, May 19, 2016, https:// chicago.eater.com/2016/5/19/11695724/alinea-chicago-grant-achatz-nick -kokonas; Noah Kagan, "Lessons From the World's Best Restaurant," OkDork (blog), March 15, 2019, https://okdork.com/lessons-worlds-best-restaurant; "No. 1: Alinea," Best Restaurants in Chicago, *Chicago Magazine*, July 2018, www .chicagomag.com/dining-drinking/July-2018/The-50-Best-Restaurants-in -Chicago/Alinea.

29. "No. 1: Alinea."

30. Robert M. Pirsig, *Zen and the Art of Motorcycle Maintenance: An Inquiry into Values* (New York: Morrow, 1984), 88.

31. Emma Court, "Who Is Merck CEO Kenneth Frazier," *Business Insider*, April 17, 2019, www.businessinsider.com/who-is-merck-ceo-kenneth-frazier-2019-4.

32. Adam Grant, *Originals: How Non-Conformists Move the World* (New York: Viking, 2016).

33. Lisa Bodell, *Kill the Company: End the Status Quo, Start an Innovation Revolution* (Brookline, MA: Bibliomotion, 2016).

34. Al Pittampalli, "How Changing Your Mind Makes You a Better Leader," *Quartz*, January 25, 2016, https://qz.com/598998/how-changing-your-mind -makes-you-a-better-leader.

35. David Mikkelson, "NASA's 'Astronaut Pen,'" *Snopes*, April 19, 2014, www .snopes.com/fact-check/the-write-stuff.

36. Albert Einstein, *On the Method of Theoretical Physics* (New York, Oxford University Press, 1933).

37. Carl Sagan, *The Demon-Haunted World: Science as a Candle in the Dark* (New York: Random House, 1995; repr., Ballantine, 1997), 211.

38. TVTropes, "Occam's Razor," https://tvtropes.org/pmwiki/pmwiki.php /Main/OccamsRazor.

39. David Kord Murray, *Borrowing Brilliance: The Six Steps to Business Innovation by Building on the Ideas of Others* (New York: Gotham Books, 2009).

40. Peter Attia, interview with author, August 2018.

41. Mary Roach, *Packing for Mars: The Curious Science of Life in the Void* (New York: W.W. Norton, 2010), 189.

42. Chaikin, "Changing the Rocket Equation?"

43. Fernholz, *Rocket Billionaires*, 83.

44. Fernholz, *Rocket Billionaires*, 83.

45. Chris Hadfield, *An Astronaut's Guide to Life on Earth: What Going to Space Taught Me About Ingenuity, Determination, and Being Prepared for Anything* (New York: Little, Brown and Company, 2013).

46. Richard Hollingham, "Soyuz: The Soviet Space Survivor," *BBC Future*, December 2, 2014, www.bbc.com/future/story/20141202-the-greatest-spacecraft-ever.

47. E. F. Schumacher, *Small Is Beautiful: Economics As If People Mattered* (New York: Harper Perennial, 2010).

48. Kyle Stock, "The Little Ion Engine That Could," *Bloomberg Businessweek*, July 26, 2018, www.bloomberg.com/news/features/2018-07-26/ion-engine-startup -wants-to-change-the-economics-of-earth-orbit.

49. Stock, "Little Ion Engine."

50. Tracy Staedter, "Dime-Size Thrusters Could Propel Satellites, Spacecraft," *Space.com*, March 23, 2017, www.space.com/36180-dime-size-accion-thrusters -propel-spacecraft.html.

51. Keith Tidman, "Occam's Razor: On the Virtue of Simplicity," *Philosophical Investigations*, May 28, 2018, www.philosophical-investigations.org/2018/05 /occams-razor-on-virtue-of-simplicity.html.

52. Sarah Freeman, "Alinea 2.0: Reinventing One of the World's Best Restaurants: Why Grant Achatz and Nick Kokonas Hit the Reset Button," *Chicago Eater*, May 19, 2016, https://chicago.eater.com/2016/5/19/11695724/alinea -chicago-grant-achatz-nick-kokonas.

53. Richard Duppa et al., *The Lives and Works of Michael Angelo and Raphael* (London: Bell & Daldy 1872), 151.

54. Jeffrey H. Dyer, Hal Gregersen, and Clayton M. Christensen, "The Innovator's DNA," *Harvard Business Review*, December 2009, https://hbr.org /2009/12/the-innovators-dna.

55. H. L. Mencken, *Prejudices: Second Series* (London: Jonathan Cape, 1921), 158, https://archive.org/details/prejudicessecond00mencuoft/page/158.

56. Alfred North Whitehead, *The Concept of Nature: Tarner Lectures Delivered in Trinity College* (Cambridge: University Press, 1920), 163.

Chapter 3: A Mind at Play

1. The opening section on Albert Einstein's thought experiments is based on the following sources: Walter Isaacson, "The Light-Beam Rider," *New York Times*, October 30, 2015; Albert Einstein, "Albert Einstein: Notes for an Auto-

biography," *Saturday Review*, November 26, 1949, https://archive.org/details /EinsteinAutobiography; Walter Isaacson, *Einstein: His Life and Universe* (New York: Simon & Schuster, 2007); Albert Einstein, *The Collected Papers of Albert Einstein*, vol. 7, *The Berlin Years: Writings, 1918–1921* (English translation supplement), trans. Alfred Engel (Princeton, NJ: Princeton University, 2002), https:// einsteinpapers.press.princeton.edu/vol7-trans/152; Kent A. Peacock, "Happiest Thoughts: Great Thought Experiments in Modern Physics," in *The Routledge Companion to Thought Experiments*, ed. Michael T. Stuart, Yiftach Fehige, and James Robert Brown, Routledge Philosophy Companions (London and New York: Routledge/Taylor & Francis Group, 2018).

2. Isaacson, *Einstein: His Life and Universe*, 27.

3. Letitia Meynell, "Images and Imagination in Thought Experiments," in *The Routledge Companion to Thought Experiments*, ed. Michael T. Stuart, Yiftach Fehige, and James Robert Brown, Routledge Philosophy Companions (London and New York: Routledge/Taylor & Francis Group, 2017) (internal quotation marks omitted).

4. James Robert Brown, *The Laboratory of the Mind: Thought Experiments in the Natural Sciences* (New York: Routledge, 1991; reprint 2005).

5. John J. O'Neill, *Prodigal Genius: The Life of Nikola Tesla* (New York: Cosimo, 2006), 257.

6. Nikola Tesla, *My Inventions: The Autobiography of Nikola Tesla* (New York: Penguin, 2011).

7. Walter Isaacson, *Leonardo da Vinci* (New York: Simon & Schuster, 2017), 196.

8. Albert Einstein, *Ideas and Opinions* (New York: Bonanza Books, 1954), 274.

9. Shane Parrish, "Thought Experiment: How Einstein Solved Difficult Problems," Farnam Street (blog), June 2017, https://fs.blog/2017/06/thought -experiment-how-einstein-solved-difficult-problems.

10. NASA, "The Apollo 15 Hammer-Feather Drop," February 11, 2016, https:// nssdc.gsfc.nasa.gov/planetary/lunar/apollo_15_feather_drop.html.

11. Rachel Feltman, "Schrödinger's Cat Just Got Even Weirder (and Even More Confusing)," *Washington Post*, May 27, 2016, www.washingtonpost.com /news/speaking-of-science/wp/2016/05/27/schrodingers-cat-just-got-even -weirder-and-even-more-confusing/?utm_term=.ed0e9088a988.

12. Sergey Armeyskov, "Decoding #RussianProverbs: Proverbs With the Word 'Nos[e],'" Russian Universe (blog), December 1, 2014, https://russianuniverse .org/2014/01/12/russian-saying-2/#more-1830.

13. Brian Grazer and Charles Fishman, *A Curious Mind: The Secret to a Bigger Life* (New York: Simon & Shuster, 2015; reprinted 2016), 11.

14. Todd B. Kashdan, "Companies Value Curiosity but Stifle It Anyway," *Harvard Business Review*, October 21, 2015, https://hbr.org/2015/10/companies -value-curiosity-but-stifle-it-anyway.

15. George Bernard Shaw, Quotable Quotes, *Reader's Digest*, May 1933, 16.

16. The discussion of instant photography is based on Christopher Bonanos, *Instant: The Story of Polaroid* (New York: Princeton Architectural Press, 2012), 32; Warren Berger, *A More Beautiful Question: The Power of Inquiry to Spark Breakthrough Ideas* (New York: Bloomsbury USA, 2014), 72–73; American Chemical Society, "Invention of Polaroid Instant Photography," www.acs.org/content/acs /en/education/whatischemistry/landmarks/land-instant-photography.html #invention_of_instant_photography.

17. Jennifer Ludden, "The Appeal of 'Harold and the Purple Crayon,'" NPR, May 29, 2005, www.npr.org/templates/story/story.php?storyId=4671937.

18. Peter Galison, *Einstein's Clocks, Poincaré's Maps: Empires of Time* (New York: W.W. Norton, 2003).

19. Isaacson, *Leonardo Da Vinci*, 520.

20. David Brewster, *Memoirs of the Life, Writings, and Discoveries of Sir Isaac Newton* (Edinburgh: Thomas Constable and Co., 1855), 407.

21. James March, "Technology of Foolishness," first published in *Civiløkonomen* (Copenhagen, 1971), www.creatingquality.org/Portals/1/DNNArticleFiles/6346 31045269246454the%20technology%20of%20foolishness.pdf.

22. The research is summarized in Darya L. Zabelina and Michael D. Robinson, "Child's Play: Facilitating the Originality of Creative Output by a Priming Manipulation," *Psychology of Aesthetics, Creativity, and the Arts* 4, no. 1 (2010): 57–65, www.psychologytoday.com/files/attachments/34246/zabelina-robinson -2010a.pdf.

23. Zabelina and Robinson, "Child's Play."

24. Massachusetts Institute of Technology, "The MIT Press and the MIT Media Lab Launch the Knowledge Futures Group," press release, September 25, 2018, https://mitpress.mit.edu/press-news/Knowledge-Futures-Group-launch; MIT Media Lab, "Lifelong Kindergarten: Engaging People in Creative Learning Experiences," press release, www.media.mit.edu/groups/lifelong-kindergarten /overview.

25. Isaacson, *Leonardo da Vinci*, 353–354.

26. Bureau of Labor Statistics, "American Time Use Survey," 2017, www.bls .gov/tus/a1_2017.pdf.

27. Timothy D. Wilson et al., "Just Think: The Challenges of the Disengaged Mind," *Science*, February 17, 2015, www.ncbi.nlm.nih.gov/pmc/articles/PMC 4330241.

28. Edward O. Wilson, *Consilience: The Unity of Knowledge* (New York: Alfred A. Knopf, 1998), 294.

29. William Deresiewicz, lecture at US Military Academy, West Point, October 2009; subsequently published as an essay: William Deresiewicz, "Solitude and Leadership," *American Scholar*, March 1, 2010.

30. Teresa Belton and Esther Priyadharshini, "Boredom and Schooling: A Cross-Disciplinary Exploration," *Cambridge Journal of Education*, December 1, 2007, www.ingentaconnect.com/content/routledg/ccje/2007/00000037/00000 004/art00008.

31. Taki Takeuchi et al., "The Association Between Resting Functional Connectivity and Creativity," *Cerebral Cortex* 22 (2012): 2921–2929; Simone Kühn et al. "The Importance of the Default Mode Network in Structural MRI Study," *Journal of Creative Behavior* 48 (2014): 152–163, www.researchgate.net/publication /259539395_The_Importance_of_the_Default_Mode_Network_in_Creativity -A_Structural_MRI_Study; James Danckert and Colleen Merrifield, "Boredom, Sustained Attention and the Default Mode Network," *Experimental Brain Research* 236, no. 9 (2016), www.researchgate.net/publication/298739805_Boredom _sustained_attention_and_the_default_mode_network.

32. David Kord Murray, *Borrowing Brilliance: The Six Steps to Business Innovation by Building on the Ideas of Others* (New York: Gotham Books, 2009).

33. Benedict Carey, "You're Bored, but Your Brain Is Tuned In," *New York Times*, August 5, 2008, www.nytimes.com/2008/08/05/health/research/05mind .html.

34. Alex Soojung-Kim Pang, *Rest: Why You Get More Done When You Work Less* (New York: Basic Books, 2016), 100.

35. David Eagleman, *The Brain: The Story of You* (Edinburgh, UK: Canongate Books, 2015).

36. Edwina Portocarrero, David Cranor, and V. Michael Bove, "Pillow-Talk: Seamless Interface for Dream Priming Recalling and Playback," Massachusetts Institute of Technology, 2011, http://web.media.mit.edu/~vmb/papers/4p375 -portocarrero.pdf.

37. David Biello, "Fact or Fiction? Archimedes Coined the Term 'Eureka!' in the Bath," *Scientific American*, December 8, 2006, www.scientificamerican.com /article/fact-or-fiction-archimede/?redirect=1.

38. "Ken952," "Office Shower," video, YouTube, uploaded August 23, 2008, www.youtube.com/watch?v=dHG_bjGschs.

39. "Idea For Hubble Repair Device Born in the Shower," *Baltimore Sun*, November 30, 1993, www.baltimoresun.com/news/bs-xpm-1993-11-30-1993334170 -story.html.

40. Denise J. Cai et al., "REM, Not Incubation, Improves Creativity by Priming Associative Networks," *Proceedings of the National Academy of Sciences* 106, no. 25 (June 23, 2009): 10,130–10,134, www.pnas.org/content/106/25/10130.full.

41. Ben Orlin, "The State of Being Stuck," Math With Bad Drawings (blog), September 20, 2017, https://mathwithbaddrawings.com/2017/09/20/the-state-of -being-stuck.

42. NOVA, "Solving Fermat: Andrew Wiles," interview with Andrew Wiles, PBS, October 31, 2000, www.pbs.org/wgbh/nova/article/andrew-wiles-fermat.

43. Judah Pollack and Olivia Fox Cabane, *Butterfly and the Net: The Art and Practice of Breakthrough Thinking* (New York: Portfolio/Penguin, 2017), 44–45.

44. Cameron Prince, "Nikola Tesla Timeline," Tesla Universe, https://tesla universe.com/nikola-tesla/timeline/1882-tesla-has-ac-epiphany.

45. Damon Young, "Charles Darwin's Daily Walks," *Psychology Today*, January 12, 2015, www.psychologytoday.com/us/blog/how-think-about-exercise/201 501/charles-darwins-daily-walks.

46. Pang, *Rest*, 100.

47. Melissa A. Schilling, *Quirky: The Remarkable Story of the Traits, Foibles, and Genius of Breakthrough Innovators Who Changed the World* (New York: Public Affairs, 2018).

48. Cal Newport, "Neil Gaiman's Advice to Writers: Get Bored," Cal Newport website, November 11, 2016, www.calnewport.com/blog/2016/11/11/neil-gaimans -advice-to-writers-get-bored.

49. Stephen King, *On Writing: A Memoir of the Craft* (New York: Scribner, 2000).

50. Mo Gawdat, *Solve for Happy: Engineering Your Path to Joy* (New York: North Star Way, 2017), 118.

51. Rebecca Muller, "Bill Gates Spends Two Weeks Alone in the Forest Each Year. Here's Why," *Thrive Global*, July 23, 2018, https://thriveglobal.com/stories /bill-gates-think-week.

52. Phil Knight, *Shoe Dog: A Memoir by the Creator of Nike* (New York: Scribner, 2016).

53. Rainer Maria Rilke, *Letters to a Young Poet* (New York: Penguin, 2012), 21.

54. Scott A. Sandford, "Apples and Oranges: A Comparison," *Improbable Research* (1995), www.improbable.com/airchives/paperair/volume1/v1i3/air-1-3 -apples.php.

55. Waqas Ahmed, *The Polymath: Unlocking the Power of Human Versatility* (West Sussex, UK: John Wiley & Sons, 2018).

56. Andrew Hill, "The Hidden Benefits of Hiring Jacks and Jills of All Trades," *Financial Times*, February 10, 2019, www.ft.com/content/e7487264-2ac0-11e9 -88a4-c32129756dd8.

57. Jaclyn Gurwin et al., "A Randomized Controlled Study of Art Observation Training to Improve Medical Student Ophthalmology Skills," *Ophthalmology* 125, no. 1 (January 2018): 8–14, www.ncbi.nlm.nih.gov/pubmed/28781219.

58. John Murphy, "Medical School Won't Teach You to Observe—But Art Class Will, Study Finds," *MDLinx*, September 8, 2017, www.mdlinx.com /internal-medicine/article/1101 (emphasis in original).

59. François Jacob, "Evolution and Tinkering," *Science*, June 10, 1977.

60. Gary Wolf, "Steve Jobs: The Next Insanely Great Thing," *Wired*, February 1, 1996, www.wired.com/1996/02/jobs-2.

61. Albert Einstein, *Ideas and Opinions: Based on Mein Weltbild* (New York: Crown, 1954).

62. P. W. Anderson, "More Is Different," *Science*, August 4, 1972, available at www.tkm.kit.edu/downloads/TKM1_2011_more_is_different_PWA.pdf.

63. D. K. Simonton, "Foresight, Insight, Oversight, and Hindsight in Scientific Discovery: How Sighted Were Galileo's Telescopic Sightings?," *Psychology of Aesthetics, Creativity, and the Arts* (2012); Robert Kurson, *Rocket Men: The Daring Odyssey of Apollo 8 and the Astronauts Who Made Man's First Journey to the Moon* (New York: Random House, 2018).

64. Isaacson, *Leonardo Da Vinci*.

65. Sarah Knapton, "Albert Einstein's Theory of Relativity Was Inspired by Scottish Philosopher," *(London) Telegraph*, February 19, 2019, www.msn.com /en-ie/news/offbeat/albert-einsteins-theory-of-relativity-was-inspired-by -scottish-philosopher/ar-BBTMyMO.

66. Sir Charles Lyell, *Principles of Geology: Modern Changes of the Earth and Its Inhabitants* (New York: D. Appleton and Co., 1889).

67. Murray, *Borrowing Brilliance*.

68. Murray, *Borrowing Brilliance*.

69. Ryan Holiday, *Perennial Seller: The Art of Making and Marketing Work That Lasts* (New York: Portfolio/Penguin, 2017), 35; Tim Ferriss, "Rick Rubin on Cultivating World-Class Artists (Jay Z, Johnny Cash, etc.), Losing 100+ Pounds, and Breaking Down the Complex," episode 76 (podcast), *The Tim Ferriss Show*, https://tim.blog/2015/05/15/rick-rubin.

70. Matthew Braga, "The Verbasizer Was David Bowie's 1995 Lyric-Writing Mac App," *Motherboard*, January 11, 2016, https://motherboard.vice.com/en_us /article/xygxpn/the-verbasizer-was-david-bowies-1995-lyric-writing-mac-app.

71. Amy Zipkin, "Out of Africa, Onto the Web," *New York Times*, December 17, 2006, www.nytimes.com/2006/12/17/jobs/17boss.html.

72. The discussion on the Nike Waffle Trainer is based on the following sources: Knight, *Shoe Dog*; Chris Danforth, "A Brief History of Nike's Revolutionary Waffle Trainer," *Highsnobiety*, March 30, 2017, www.highsnobiety.com/2017 /03/30/nike-waffle-trainer-history; Matt Blitz, "How a Dirty Old Waffle Iron Became Nike's Holy Grail," *Popular Mechanics*, July 15, 2016, www.popular mechanics.com/technology/gadgets/a21841/nike-waffle-iron.

73. Riley Black, "Thomas Henry Huxley and the Dinobirds," *Smithsonian*, December 7, 2010, www.smithsonianmag.com/science-nature/thomas-henry-huxley -and-the-dinobirds-88519294.

74. William C. Taylor and Polly Labarre, "How Pixar Adds a New School of Thought to Disney," *New York Times*, January 29, 2006, www.nytimes.com/2006 /01/29/business/yourmoney/how-pixar-adds-a-new-school-of-thought-to -disney.html; Ed Catmull and Amy Wallace, *Creativity, Inc.: Overcoming the*

Unseen Forces That Stand in the Way of True Inspiration (Toronto: Random House Canada, 2014).

75. Frans Johansson, *The Medici Effect: What Elephants and Epidemics Can Teach Us About Innovation* (Boston: Harvard Business School Press, 2017).

76. Steve Squyres, *Roving Mars: Spirit, Opportunity, and the Exploration of the Red Planet* (New York: Hyperion, 2005); University of California Television, "Roving Mars with Steve Squyres: Conversations with History," video, YouTube, uploaded August 18, 2011, www.youtube.com/watch?v=NI6KEzsb26U&feature =youtu.be.

77. Squyres, *Roving Mars*.

78. Ethan Bernstein, Jesse Shoreb, and David Lazer, "How Intermittent Breaks in Interaction Improve Collective Intelligence," *Proceedings of the National Academy of Sciences* 115, no. 35 (August 28, 2018): 8734–8739, www.pnas.org/content /pnas/115/35/8734.full.pdf; HBS [Harvard Business School] Communications, "Problem-Solving Techniques Take On New Twist," *Harvard Gazette*, August 15, 2018, https://news.harvard.edu/gazette/story/2018/08/collaborate-on -complex-problems-but-only-intermittently.

79. Bernstein, Shoreb, and Lazer, "Intermittent Breaks."

80. Bernstein, Shoreb, and Lazer, "Intermittent Breaks."

81. Isaac Asimov, "On Creativity," 1959, first published in *MIT Technology Review*, October 20, 2014.

82. Dean Keith Simonton, *Origins of Genius: Darwinian Perspectives on Creativity* (New York: Oxford University Press, 1999), 125.

83. *Encyclopaedia Britannica*, s.v. "Alfred Wegener," updated April 5, 2019, www.britannica.com/biography/Alfred-Wegener.

84. Joseph Sant, "Alfred Wegener's Continental Drift Theory," *Scientus*, 2018, www.scientus.org/Wegener-Continental-Drift.html.

85. Mario Livio, *Brilliant Blunders: From Darwin to Einstein—Colossal Mistakes by Great Scientists That Changed Our Understanding of Life and the Universe* (New York: Simon & Schuster, 2013), 265.

86. Albert Einstein, "Zur Elektrodynamik bewegter Körper" [On the electrodynamics of moving bodies], *Annalen der Physik* 17, no. 10 (June 30, 1905).

87. Shunryu Suzuki and Richard Baker, *Zen Mind, Beginner's Mind* (Boston: Shambhala, 2006), 1.

88. Suzuki and Baker, *Zen Mind, Beginner's Mind*.

89. Alison Flood, "JK Rowling Says She Received 'Loads' of Rejections Before Harry Potter Success," *Guardian*, March 24, 2015, www.theguardian.com/books /2015/mar/24/jk-rowling-tells-fans-twitter-loads-rejections-before-harry-potter -success.

90. "Revealed: The Eight-Year-Old Girl Who Saved Harry Potter," *(London) Independent*, July 3, 2005, www.independent.co.uk/arts-entertainment/books /news/revealed-the-eight-year-old-girl-who-saved-harry-potter-296456.html.

Chapter 4: Moonshot Thinking

1. The section on Project Loon is based on the following sources: "Google Launches Product Loon," *New Zealand Herald*, June 15, 2013, www.nzherald .co.nz/internet/news/article.cfm?c_id=137&objectid=10890750; "Google Tests Out Internet-Beaming Balloons in Skies Over New Zealand," *(San Francisco) SFist*, June 16, 2013, http://sfist.com/2013/06/16/google_tests_out_internet -beaming_b.php; Derek Thompson, "Google X and the Science of Radical Creativity," *Atlantic*, November 2017, www.theatlantic.com/magazine/archive/2017 /11/x-google-moonshot-factory/540648/; Loon.com, "Loon: The Technology," video, YouTube, uploaded June 14, 2013, www.youtube.com/watch?v=mcw6j -QWGMo&feature=youtu.be; Alex Davies, "Inside X, the Moonshot Factory Racing to Build the Next Google," *Wired*, July 11, 2018, www.wired.com/story /alphabet-google-x-innovation-loon-wing-graduation; Steven Levy, "The Untold Story of Google's Quest to Bring the Internet Everywhere—by Balloon," *Wired*, August 13, 2013, www.wired.com/2013/08/googlex-project-loon.

2. Chris Anderson, "Mystery Object in Sky Captivates Locals," *Appalachian News-Express*, October 19, 2012, www.news-expressky.com/news/article_f257 128c-1979-11e2-a94e-0019bb2963f4.html.

3. Thompson, "Radical Creativity."

4. Telefónica, "Telefónica and Project Loon Collaborate to Provide Emergency Mobile Connectivity to Flooded Areas of Peru," Telefónica, May 17, 2017, www.telefonica.com/en/web/press-office/-/telefonica-and-project-loon -collaborate-to-provide-emergency-mobile-connectivity-to-flooded-areas-of -peru.

5. Alastair Westgarth, "Turning on Project Loon in Puerto Rico," *Medium*, October 20, 2017, https://medium.com/loon-for-all/turning-on-project-loon-in -puerto-rico-f3aa41ad2d7f.

6. Robert Kurson, *Rocket Men: The Daring Odyssey of Apollo 8 and the Astronauts Who Made Man's First Journey to the Moon* (New York: Random House, 2019), 17.

7. *In the Shadow of the Moon*, directed by Dave Sington (Velocity/Think Film, 2008), DVD.

8. Jade Boyd, "JFK's 1962 Moon Speech Still Appeals 50 Years Later," Rice University News, August 30, 2012, http://news.rice.edu/2012/08/30/jfks-1962 -moon-speech-still-appeals-50-years-later.

9. Gene Kranz, *Failure Is Not an Option: Mission Control from Mercury to Apollo 13 and Beyond* (New York: Simon & Schuster, 2000), 56.

10. Kranz, *Failure Is Not an Option.*

11. Mo Gawdat, *Solve for Happy: Engineering Your Path to Joy* (New York: North Star Way, 2017).

12. James Carville and Paul Begala, *Buck Up, Suck Up . . . and Come Back When You Foul Up: 12 Winning Secrets from the War Room* (New York: Simon & Schuster, 2003), 89–90.

13. Abraham Maslow, quoted in Jim Whitt, *Road Signs for Success* (Stillwater, OK: Lariat Press, 1993), 61.

14. Seth Godin, *The Icarus Deception: How High Will You Fly?* (New York: Portfolio/Penguin, 2012).

15. Shane Snow, *Smartcuts: The Breakthrough Power of Lateral Thinking* (New York: HarperBusiness, 2014), 180, Kindle.

16. Pascal-Emmanuel Gobry, "Facebook Investor Wants Flying Cars, Not 140 Characters," *Business Insider*, July 301, 2011, www.businessinsider.com/founders -fund-the-future-2011-7.

17. Jennifer Reingold, "Hondas in Space," *Fast Company*, October 5, 2005, www.fastcompany.com/74516/hondas-space-2.

18. Astro Teller, "The Head of 'X' Explains How to Make Audacity the Path of Least Resistance," *Wired*, April 15, 2016, www.wired.com/2016/04/the-head -of-x-explains-how-to-make-audacity-the-path-of-least-resistance/#.2vy7nkes6.

19. Lisa Bodell, *Kill the Company: End the Status Quo, Start an Innovation Revolution* (Brookline, MA: Bibliomotion, 2016), 128–129.

20. David J. Schwartz, *The Magic of Thinking Big*, Touchstone hardcover edition (New York: Touchstone, 2015), 9.

21. Dana Goodyear, "Man of Extremes: Return of James Cameron," *New Yorker*, October 19, 2009, www.newyorker.com/magazine/2009/10/26/man-of -extremes.

22. Chantal Da Silva, "Michelle Obama Tells A Secret: 'I Have Been at Every Powerful Table You Can Think Of . . . They Are Not That Smart,'" *Newsweek*, December 4, 2018, www.newsweek.com/michelle-obama-tells-secret-i-have -been-at-every-powerful-table-you-can-think-1242695.

23. On bees' learning ability, see Hamida B. Mirwan and Peter G. Kevan, "Problem Solving by Worker Bumblebees *Bombus impatiens* (Hymenoptera: Apoidea)," *Animal Cognition* 17 (September 2014): 1053–1061. On bees' ability to teach, see Kristin Hugo, "Intelligence Test Shows Bees Can Learn to Solve Tasks from Other Bees," *News Hour*, PBS, February 23, 2017, www.pbs.org/newshour /science/intelligence-test-shows-bees-can-learn-to-solve-tasks-from-other-bees.

24. Maurice Maeterlinck, *The Life of the Bee*, trans. Alfred Sutro (New York: Dodd, Mead and Company, 1915), 145–146.

25. David Deutsch, *The Beginning of Infinity: Explanations That Transform the World* (London: Allen Lane, 2011).

26. John D. Norton, "How Einstein Did Not Discover," *Physics in Perspective*, 258 (2016) www.pitt.edu/~jdnorton/papers/Einstein_Discover_final.pdf.

27. Richard W. Woodman, John E. Sawyer, and Ricky W. Griffin, "Toward a Theory of Organizational Creativity," *Academy of Management Review* 18, no. 2

(April 1993): 293; Scott David Williams, "Personality, Attitude, and Leader Influences on Divergent Thinking and Creativity in Organizations," *European Journal of Innovation Management* 7, no. 3 (September 1, 2004): 187–204; J. P. Guilford, "Cognitive Psychology's Ambiguities: Some Suggested Remedies," *Psychological Review* 89, no. 1 (1982): 48–59, https://psycnet.apa.org/record/1982 -07070-001.

28. Ting Zhang, Francesca Gino, and Joshua D. Margolis, "Does 'Could' Lead to Good? On the Road to Moral Insight," *Academy of Management Journal* 61, no. 3 (June 22, 2008), https://journals.aom.org/doi/abs/10.5465/amj.2014.0839.

29. E. J. Langer and A. I. Piper, "The Prevention of Mindlessness," *Journal of Personality and Social Psychology* 53, no. 2 (1987): 280–287.

30. Louise Lee, "Managers Are Not Always the Best Judge of Creative Ideas," *Stanford Business*, January 26, 2016, www.gsb.stanford.edu/insights/managers -are-not-best-judge-creative-ideas.

31. Justin M. Berg, "Balancing on the Creative Highwire: Forecasting the Success of Novel Ideas in Organizations," *Administrative Science Quarterly*, July 2016, www.gsb.stanford.edu/faculty-research/publications/balancing-creative -high-wire-forecasting-success-novel-ideas.

32. "Everything You Know About Genius May Be Wrong," *Heleo*, September 6, 2017, https://heleo.com/conversation-everything-know-genius-may-wrong /15062.

33. Alex Soojung-Kim Pang, *Rest: Why You Get More Done When You Work Less* (New York: Basic Books, 2016), 44.

34. Naama Mayseless, Judith Aharon-Perez, and Simone Shamay-Tsoory, "Unleashing Creativity: The Role of Left Temporoparietal Regions in Evaluation and Inhibiting the Generation of Creative Ideas," *Neuropsychologia* 64 (November 2014): 157–168.

35. I. Bernard Cohen, "Faraday and Franklin's 'Newborn Baby,'" *Proceedings of the American Philosophical Society* 131, no. 2 (June 1987): 77–182, www.jstor.org /stable/986790?read-now=1&seq=6#page_scan_tab_contents.

36. For the 2003 Mars Exploration Rovers program, see Jet Propulsion Laboratory, California Institute of Technology, "Spacecraft: Airbags," NASA, https:// mars.nasa.gov/mer/mission/spacecraft_edl_airbags.html. For 2008 Phoenix mission, see NASA, "NASA Phoenix Mission Ready for Mars Landing," May 13, 2008, press release, www.nasa.gov/mission_pages/phoenix/news/phoenix-2008 050813.html.

37. Adam Steltzner and William Patrick, *Right Kind of Crazy: A True Story of Teamwork, Leadership, and High-Stakes Innovation* (New York: Portfolio/Penguin, 2016), 137.

38. Arnold Schwarzenegger, with Peter Petre, *Total Recall: My Unbelievably True Life Story* (New York: Simon & Schuster, 2012), 53.

39. Arnold Schwarzenegger, "Shock Me," Arnold Schwarzenegger website, July 30, 2012, www.schwarzenegger.com/fitness/post/shock-me.

40. Bernard D. Beitman, "Brains Seek Patterns in Coincidences," *Psychiatric Annals* 39, no. 5 (May 2009): 255–264, https://drjudithorloff.com/main/wp-content/uploads/2017/09/Psychiatric-Annals-Brains-Seek-Patterns.pdf.

41. Norman Doidge, *The Brain's Way of Healing: Remarkable Discoveries and Recoveries from the Frontiers of Neuroplasticity* (New York: Penguin Books, 2015).

42. Paul J. Steinhardt, "What Impossible Meant to Feynman," *Nautilus*, January 31, 2019, http://m.nautil.us/issue/68/context/what-impossible-meant-to-feynman.

43. Alok Jha, "Science Weekly with Michio Kaku: Impossibility Is Relative," *Guardian* (US edition), June 14, 2009, www.theguardian.com/science/audio/2009/jun/11/michio-kaku-physics-impossible.

44. Andrea Estrada, "Reading Kafka Improves Learning, Suggests UCSB Psychology Study," *UC Santa Barbara Current*, September 15, 2009, www.news.ucsb.edu/2009/012685/reading-kafka-improves-learning-suggests-ucsb-psychology-study.

45. Adam Morgan and Mark Barden, *A Beautiful Constraint: How to Transform Your Limitations into Advantages, and Why It's Everyone's Business* (Hoboken, NJ: Wiley, 2015).

46. Travis Proulx and Steven J. Heine, "Connections from Kafka: Exposure to Meaning Threats Improves Implicit Learning of an Artificial Grammar," *Psychological Science* 20, no. 9 (2009): 1125–1131.

47. Bill Ryan, "What Verne Imagined, Sikorsky Made Fly," *New York Times*, May 7, 1995, www.nytimes.com/1995/05/07/nyregion/what-verne-imagined-sikorsky-made-fly.html.

48. Mark Strauss, "Ten Inventions Inspired by Science Fiction," *Smithsonian Magazine*, March 15, 2012, www.smithsonianmag.com/science-nature/ten-inventions-inspired-by-science-fiction-128080674.

49. Tim Fernholz, *Rocket Billionaires: Elon Musk, Jeff Bezos, and the New Space Race* (Boston: Houghton Mifflin Harcourt, 2018), 69.

50. Dylan Minor, Paul Brook, and Josh Bernoff, "Data From 3.5 Million Employees Shows How Innovation Really Works," *Harvard Business Review*, October 9, 2017, https://hbr.org/2017/10/data-from-3-5-million-employees-shows-how-innovation-really-works.

51. Neil Strauss, "Elon Musk: The Architect of Tomorrow," *Rolling Stone*, November 15, 2017, www.rollingstone.com/culture/culture-features/elon-musk-the-architect-of-tomorrow-120850.

52. Snow, *Smartcuts*.

53. Tom Junod, "Elon Musk: Triumph of His Will," *Esquire*, November 15, 2012, www.esquire.com/news-politics/a16681/elon-musk-interview-1212.

54. Michael Belfiore, "Behind the Scenes with the World's Most Ambitious Rocket Makers," *Popular Mechanics*, September 1, 2009, www.popularmechanics .com/space/rockets/a5073/4328638.

55. Junod, "Musk: Triumph of His Will."

56. Andrew Chaikin, "Is SpaceX Changing the Rocket Equation?," *Smithsonian*, January 2012, www.airspacemag.com/space/is-spacex-changing-the-rocket -equation-132285884/?no-ist=&page=2.

57. Sam Altman, "How to Be Successful," Sam Altman (blog), January 24, 2019, http://blog.samaltman.com/how-to-be-successful.

58. X, "Obi Felten, Head of Getting Moonshots Ready for Contact with the Real World," https://x.company/team/obi.

59. Davies, "Inside X, the Moonshot Factory."

60. Thompson, "Radical Creativity."

61. Jessica Guynn, "Google's Larry Page Will Try to Recapture Original Energy as CEO," *Los Angeles Times*, January 22, 2011, www.latimes.com/business /la-xpm-2011-jan-22-la-fi-google-20110122-story.html.

62. Leah Binkovitz, "Tesla at the Smithsonian: The Story Behind His Genius," *Smithsonian*, June 27, 2013, www.smithsonianmag.com/smithsonian-institution /tesla-at-the-smithsonian-the-story-behind-his-genius-3329176; Jill Jonnes, *Empires of Light: Edison, Tesla, Westinghouse, and the Race to Electrify the World* (New York: Random House, 2003).

63. Obi Felten, "Watching Loon and Wing Grow Up," LinkedIn, August 1, 2018, www.linkedin.com/pulse/watching-loon-wing-grow-up-obi-felten.

64. Obi Felten, interview with author, July 2019.

65. Obi Felten, "Living in Modern Times: Why We Worry About New Technology and What We Can Do About It," LinkedIn, January 12, 2018, www .linkedin.com/pulse/living-modern-times-why-we-worry-new-technology-what -can-obi-felten.

66. Astro Teller, "The Secret to Moonshots? Killing Our Projects," *Wired*, February 16, 2016, www.wired.com/2016/02/the-secret-to-moonshots-killing-our -projects/#.euwa8vwaq.

67. Astro Teller, "The Head of 'X' Explains How to Make Audacity the Path of Least Resistance," *Wired*, April 15, 2016, www.wired.com/2016/04/the-head -of-x-explains-how-to-make-audacity-the-path-of-least-resistance/#.2vy7nkes6.

68. Davies, "Inside X, the Moonshot Factory."

69. Thompson, "Radical Creativity"; Obi Felten, "How to Kill Good Things to Make Room for Truly Great Ones," X (blog), March 8, 2016, https://blog.x .company/how-to-kill-good-things-to-make-room-for-truly-great-ones-867 fb6ef026; Davies, "Inside X, the Moonshot Factory."

70. Thompson, "Radical Creativity."

71. Felten, "How to Kill Good Things."

72. Steven Levey, "The Untold Story of Google's Quest to Bring the Internet Everywhere—By Balloon," *Wired*, August 13, 2013, www.wired.com/2013/08/googlex-project-loon.

73. Chautauqua Institution, "Obi Felten: Head of Getting Moonshots Ready for Contact with the Real World, X," video, YouTube, uploaded June 30, 2017, www.youtube.com/watch?v=PotKc56xYyg&feature=youtu.be.

74. Mark Holmes, "It All Started with a Suit: The Story Behind Shotwell's Rise to SpaceX, *Via Satellite*, April 21, 2014, www.satellitetoday.com/business/2014/04/21/it-all-started-with-a-suit-the-story-behind-shotwells-rise-to-spacex.

75. Max Chafkin and Dana Hull, "SpaceX's Secret Weapon Is Gwynne Shotwell," *Bloomberg Businessweek*, July 26, 2018, www.bloomberg.com/news/features/2018-07-26/she-launches-spaceships-sells-rockets-and-deals-with-elon-musk.

76. Eric Ralph, "SpaceX to Leverage Boring Co. Tunneling Tech to Help Humans Settle Mars," Teslarati, May 23, 2018, www.teslarati.com/spacex-use-boring-company-tunneling-technology-mars; CNBC, "SpaceX President Gwynne Shotwell on Elon Musk and the Future of Space Launches," video, YouTube, uploaded May 22, 2018, https://youtu.be/clhXVdjvOyk.

77. The discussion on the Boring Company is based on the following sources: Boring Company, "FAQ," www.boringcompany.com/faq; Elon Musk, "The Future We're Building—and Boring," TED talk, April 2017, www.ted.com/talks/elon_musk_the_future_we_re_building_and_boring.

78. *Back to the Future*, by Robert Zemeckis and Bob Gale and directed by Robert Zemeckis (Universal Pictures, 1985). The quote was uttered by the character Emmet "Doc" Brown as he and his friends prepare to blast off to another time-traveling adventure.

79. Laura Bliss, "Dig Your Crazy Tunnel, Elon Musk!," *City Lab*, December 19, 2018, www.citylab.com/transportation/2018/12/elon-musk-tunnel-ride-tesla-boring-company-los-angeles/578536.

80. Boring Company, "Chicago," www.boringcompany.com/chicago.

81. Boring Company, "Las Vegas," www.boringcompany.com/lvcc.

82. Antoine de Saint-Exupéry, *The Wisdom of the Sands* (New York: Harcourt, Brace and Company, 1950), 155.

83. "Alan Kay, Educator and Computing Pioneer," TED speaker personal profile, March 2008, www.ted.com/speakers/alan_kay.

84. The discussion of Amazon's use of backcasting draws on the following sources: Jeff Dyer and Hal Gregersen, "How Does Amazon Stay at Day One?," *Forbes*, August 8, 2017, www.forbes.com/sites/innovatorsdna/2017/08/08/how-does-amazon-stay-at-day-one/#62a21bb67e4d; Ian McAllister, answer to submitted question: "What Is Amazon's Approach to Product Development and Product Management?," *Quora*, May 18, 2012, www.quora.com/What-is-Amazons-approach-to-product-development-and-product-management; Natalie Berg and

Miya Knights, *Amazon: How the World's Most Relentless Retailer Will Continue to Revolutionize Commerce* (New York: Kogan Page, 2019), 10.

85. Derek Sivers, "Detailed Dreams Blind You to New Means," Derek Sivers website, March 18, 2018, https://sivers.org/details.

86. Astro Teller, "Tackle the Monkey First," X, the Moonshot Factory, December 7, 2016, https://blog.x.company/tackle-the-monkey-first-90fd6223e04d.

87. Thompson, "Radical Creativity."

88. Kathy Hannun, "Three Things I Learned from Turning Seawater into Fuel," X, the Moonshot Factory, December 7, 2016, https://blog.x.company /three-things-i-learned-from-turning-seawater-into-fuel-66aeec36cfaa.

89. The discussion on Project Foghorn is based on the following sources: Hannun, "Turning Seawater into Fuel"; Teller, "Tackle the Monkey First"; Thompson, "Radical Creativity."

90. George Bernard Shaw, *Man and Superman* (Westminster: Archibald Constable & Co., 1903), 238.

91. Burt Rutan, quoted in Peter Diamandis, "True Breakthroughs = Crazy Ideas + Passion," *Tech Blog*, May 2017, www.diamandis.com/blog/true -breakthroughs-crazy-ideas-passion.

Chapter 5: What If We Sent Two Rovers Instead of One?

1. The description of landing on Mars is based on the following sources: Steve Squyres, *Roving Mars: Spirit, Opportunity, and the Exploration of the Red Planet* (New York: Hyperion, 2005), 79–80; Adam Steltzner and William Patrick, *Right Kind of Crazy: A True Story of Teamwork, Leadership, and High-Stakes Innovation* (New York: Portfolio/Penguin, 2016); Jet Propulsion Laboratory, California Institute of Technology, "Spacecraft: Aeroshell," NASA, https://mars.nasa.gov /mer/mission/spacecraft_edl_aeroshell.html; NASA Jet Propulsion Laboratory, California Institute of Technology, "Spacecraft: Aeroshell—RAD Rockets," https://mars.nasa.gov/mer/mission/spacecraft_edl_radrockets.html; Integrated Teaching and Learning Program, College of Engineering, University of Colorado Boulder, "Lesson: Six Minutes of Terror," Teach Engineering, July 31, 2017, www.teachengineering.org/lessons/view/cub_mars_lesson05.

2. Amar Toor, "NASA Details Curiosity's Mars Landing in 'Seven Minutes of Terror' Video," *Verge*, June 26, 2012, www.theverge.com/2012/6/26/3117662 /nasa-mars-rover-curiosity-seven-minutes-terror-video.

3. For distance, see NASA, "Mars Close Approach to Earth: July 31, 2018," NASA, https://mars.nasa.gov/allaboutmars/nightsky/mars-close-approach; Tim Sharp, "How Far Away Is Mars?," *Space.com*, December 15, 2017, www.space .com/16875-how-far-away-is-mars.html. For Mars's revolution speed, see NASA, "Mars Facts," NASA, https://mars.nasa.gov/allaboutmars/facts/#?c=inspace&s =distance.

4. John Maynard Keynes, *The General Theory of Employment, Interest, and Money* (New York: Harcourt, Brace, 1936).

5. Dan Meyer, "Rough-Draft Thinking & Bucky the Badger," dy/dan (blog), May 21, 2018, https://blog.mrmeyer.com/2018/rough-draft-thinking-bucky-the-badger.

6. Thomas Wedell-Wedellsborg, "Are You Solving the Right Problems?," *Harvard Business Review*, February 2017, https://hbr.org/2017/01/are-you-solving-the-right-problems.

7. Paul C. Nutt, "Surprising but True: Half the Decisions in Organizations Fail," *Academy of Management Executive* 13, no. 4 (November 1999): 75–90.

8. Nutt, "Surprising but True."

9. Merim Bilalić, Peter McLeod, and Fernand Gobet, "Why Good Thoughts Block Better Ones: The Mechanism of the Pernicious Einstellung (Set) Effect," *Cognition* 108, no. 3 (September 2008): 652–661, https://bura.brunel.ac.uk/bitstream/2438/2276/1/Einstellung-Cognition.pdf.

10. NASA, Step-by-Step Guide to Entry, Descent, and Landing https://mars.nasa.gov/mer/mission/tl_entry1.html.

11. Hal Gregersen, "Bursting the CEO Bubble," *Harvard Business Review*, April 2017, https://hbr.org/2017/03/bursting-the-ceo-bubble.

12. Charles Darwin, *The Correspondence of Charles Darwin: 1858–1859*, ed. Frederick Burkhardt and Sydney Smith (New York: Cambridge University Press, 1985).

13. Werner Heisenberg, *Physics and Philosophy: The Revolution in Modern Science* (New York: Harper, 1958).

14. Ahmed M. Abdulla et al., "Problem Finding and Creativity: A Meta-Analytic Review," *Psychology of Aesthetics, Creativity, and the Arts* (August 9, 2018), https://psycnet.apa.org/record/2018-38514-001.

15. Jacob W. Getzels and Mihaly Csikszentmihalyi, *The Creative Vision: Longitudinal Study of Problem Finding in Art* (New York: Wiley, 1976).

16. NASA, "Mariner Space Probes," https://history.nasa.gov/mariner.html.

17. NASA, "Viking 1 and 2," https://mars.nasa.gov/programmissions/missions/past/viking.

18. NASA, "Viking Mission Overview," www.nasa.gov/redplanet/viking.html.

19. Squyres, *Roving Mars*.

20. Squyres, *Roving Mars*, 90.

21. NASA, Girl with Dreams Names Mars Rovers "Spirit" and "Opportunity," (June 8, 2003) www.nasa.gov/missions/highlights/mars_rover_names.html.

22. Squyres, *Roving Mars*, 145.

23. Squyres, *Roving Mars*, 122.

24. The description of *Spirit* and *Opportunity*'s landing on Mars is based largely on Squyres, *Roving Mars*; University of California Television, "Roving Mars with

Steve Squyres: Conversations with History," video, YouTube, uploaded August 18, 2011, www.youtube.com/watch?v=NI6KEzsb26U&feature=youtu.be.

25. John Callas, "A Heartfelt Goodbye to a Spirited Mars Rover," NASA, May 25, 2011, https://mars.nasa.gov/news/1129/a-heartfelt-goodbye-to-a-spirited-mars-rover.

26. NASA, "NASA's Record-Setting Opportunity Rover Mission on Mars Comes to End," press release, February 13, 2019, www.nasa.gov/press-release/nasas-record-setting-opportunity-rover-mission-on-mars-comes-to-end.

27. World Health Organization, "Preterm Birth," February 19, 2018, www.who.int/en/news-room/fact-sheets/detail/preterm-birth.

28. Cheryl Bird, "How an Incubator Works in the Neonatal ICU," *Verywell Family*, November 6, 2018, www.verywellfamily.com/what-is-an-incubator-for-premature-infants-2748445.

29. Bird, "Neonatal ICU"; Kelsey Andeway, "Why Are Incubators Important for Babies in the NICU?," *Health eNews*, July 23, 2018, www.ahchealthenews.com/2018/07/23/incubators-important-babies-nicu.

30. Elizabeth A. Reedy, "Care of Premature Infants," University of Pennsylvania School of Nursing, www.nursing.upenn.edu/nhhc/nurses-institutions-caring/care-of-premature-infants; Vinnie DeFrancesco, "Neonatal Incubator—Perinatology," *ScienceDirect*, 2004, www.sciencedirect.com/topics/nursing-and-health-professions/neonatal-incubator.

31. The discussion of the Embrace infant warmer is based on the following sources: Snow, *Smartcuts*; Adam Morgan and Mark Barden, *A Beautiful Constraint: How to Transform Your Limitations into Advantages, and Why It's Everyone's Business* (Hoboken, NJ: Wiley, 2015); Embrace home page, www.embraceinnovations.com.

32. Stanford University, "Design for Extreme Affordability—About," https://extreme.stanford.edu/about-extreme.

33. Neil Gaiman, *The Sandman*, vol. 2, *The Doll's House*, 30th anniv. ed., issues 9–16 (Burbank, CA: DC Comics, 2018).

34. Peter Attia, interview with author, August 2018.

35. Tina Seelig, "The $5 Challenge!," *Psychology Today*, August 5, 2009, www.psychologytoday.com/us/blog/creativityrulz/200908/the-5-challenge.

36. Alexander Calandra, "Angels on a Pin," *Saturday Review*, December 21, 1968. The story also appeared in *Quick Takes: Short Model Essays for Basic Composition*, ed. Elizabeth Penfield and Theodora Hill (New York: HarperCollins College Publishers, 1995), and can be found at https://kaushikghose.files.wordpress.com/2015/07/angels-on-a-pin.pdf.

37. Robert E. Adamson, "Functional Fixedness as Related to Problem Solving: A Repetition of Three Experiments," *Journal of Experimental Psychology* 44, no. 4 (October 1952): 288–291, www.dtic.mil/dtic/tr/fulltext/u2/006119.pdf.

38. Will Yakowicz, "This Space-Age Blanket Startup Has Helped Save 200,000 Babies (and Counting)," *Inc.*, May 2016, www.inc.com/magazine/201605/will -yakowicz/embrace-premature-baby-blanket.html.

39. Patrick J. Gallagher, "Velcro," International Trademark Association, April 1, 2004, www.inta.org/INTABulletin/Pages/VELCRO.aspx.

40. Tony McCaffrey, "Innovation Relies on the Obscure: A Key to Overcoming the Classic Problem of Functional Fixedness," *Psychological Science* 23, no. 3 (February 7, 2012): 215–218, https://journals.sagepub.com/doi/abs/10.1177/095 6797611429580.

41. Ron Miller, "How AWS Came to Be," *TechCrunch*, July 2, 2016, https:// techcrunch.com/2016/07/02/andy-jassys-brief-history-of-the-genesis-of-aws.

42. Larry Dignan, "All of Amazon's 2017 Operating Income Comes from AWS," *ZDNet*, February 1, 2017, www.zdnet.com/article/all-of-amazons-2017 -operating-income-comes-from-aws.

43. Randy Hofbauer, "Amazon-Whole Foods, 1 Year Later: 4 Grocery Experts Share Their Insights," *Progressive Grocer*, June 18, 2018, https://progressivegrocer .com/amazon-whole-foods-1-year-later-4-grocery-experts-share-their-insights.

44. NASA, "Sputnik and the Dawn of the Space Age," NASA, October 10, 2007, https://history.nasa.gov/sputnik/.

45. The discussion on the origin of the global positioning system (GPS) is based on the following sources: Steven Johnson, *Where Good Ideas Come From: The Natural History of Innovation* (New York: Riverhead Books, 2011); Robert Kurson, *Rocket Men: The Daring Odyssey of Apollo 8 and the Astronauts Who Made Man's First Journey to the Moon* (New York: Random House, 2018); William H. Guier and George C. Weiffenbach, "Genesis of Satellite Navigation," *Johns Hopkins APL Technical Digest*, 18, no. 2 (1997): 178–181, www.jhuapl.edu/Content/techdigest /pdf/V18-N02/18-02-Guier.pdf.; Alan Boyle, "Sputnik Started Space Race, Anxiety," *NBC News*, October 4, 1997, www.nbcnews.com/id/3077890/ns/technology _and_science-space/t/sputnik-started-space-race-anxiety/#.XOtOsi2ZPBI.

46. *Chicago Daily News* editorial cited in Kurson, *Rocket Men*.

47. Shane Parrish, "Inversion and the Power of Avoiding Stupidity," Farnam Street (blog), October 2013, https://fs.blog/2013/10/inversion; Ray Galkowski, "Invert, Always Invert, Margin of Safety," January 9, 2011, http://amarginofsafety .com/2011/01/09/456.

48. David Kord Murray, *Borrowing Brilliance: The Six Steps to Business Innovation by Building on the Ideas of Others* (New York: Gotham Books, 2009).

49. Murray, *Borrowing Brilliance*.

50. Warren Berger, *A More Beautiful Question: The Power of Inquiry to Spark Breakthrough Ideas* (New York: Bloomsbury USA, 2014); Patagonia, "Don't Buy This Jacket, Black Friday and the *New York Times*," November 25, 2011, www .patagonia.com/blog/2011/11/dont-buy-this-jacket-black-friday-and-the-new -york-times.

51. Patagonia, "Don't Buy This Jacket."

52. The discussion of Dick Fosbury draws on the following sources: Richard Hoffer, *Something in the Air: American Passion and Defiance in the 1968 Mexico City Olympics* (New York: Free Press, 2009); James Clear, "Olympic Medalist Dick Fosbury and the Power of Being Unconventional," James Clear (blog), https://jamesclear.com/dick-fosbury; Tom Goldman, "Dick Fosbury Turned His Back on the Bar and Made a Flop a Success," NPR, October 20, 2018, www.npr .org/2018/10/20/659025445/dick-fosbury-turned-his-back-on-the-bar-and -made-a-flop-a-success.

53. Kerry Eggers, "From Flop to Smashing High Jump Success," *Portland Tribune*, July 22, 2008, https://pamplinmedia.com/component/content/article?id =71447.

54. Rod Drury, "Why Pitching a Really Bad Idea Isn't the End of the World," *Fortune*, March 23, 2016, http://fortune.com/2016/03/22/how-to-motivate-team.

55. Gregersen, "Bursting the CEO Bubble."

Chapter 6: The Power of Flip-Flopping

1. The discussion on the Mars Climate Orbiter is based on the following sources: Steve Squyres, *Roving Mars: Spirit, Opportunity, and the Exploration of the Red Planet* (New York: Hyperion, 2005); James Oberg, "Why the Mars Probe Went off Course," IEEE Spectrum, December 1, 1999, https://spectrum.ieee.org /aerospace/robotic-exploration/why-the-mars-probe-went-off-course; Edward Euler, Steven Jolly, and H. H. "Lad" Curtis, "The Failures of the Mars Climate Orbiter and Mars Polar Lander: A Perspective from the People Involved," *American Astronautical Society*, February 2001, http://web.mit.edu/16.070/www /readings/Failures_MCO_MPL.pdf; "Mars Climate Orbiter Mishap Investigation Board Phase I Report," NASA, November 10, 1999, https://llis.nasa.gov /llis_lib/pdf/1009464main1_0641-mr.pdf; House Committee on Science, Space, and Technology, "Testimony of Thomas Young, Chairman of the Mars Program Independent Assessment Team Before the House Science Committee," press release, SpaceRef, April 12, 2000, www.spaceref.com/news/viewpr.html?pid=1444.

2. NASA, "Mars Facts," https://mars.nasa.gov/allaboutmars/facts/#?c=in space&s=distance; Kathryn Mersmann, "The Fact and Fiction of Martian Dust Storms," NASA, September 18, 2015, www.nasa.gov/feature/goddard/the-fact -and-fiction-of-martian-dust-storms.

3. NASA Jet Propulsion Laboratory, "NASA's Mars Climate Orbiter Believed to Be Lost," NASA, September 23, 1999, www.jpl.nasa.gov/news/news .php?feature=5000.

4. Robert M. Pirsig, *Zen and the Art of Motorcycle Maintenance: An Inquiry into Values* (New York: Morrow, 1984), 6.

5. Jeremy A. Frimer, Linda J. Skitka, and Matt Motyl, "Liberals and Conservatives Are Similarly Motivated to Avoid Exposure to One Another's Opinions," *Journal of Experimental Social Psychology* 72 (September 2017): 1–12, www.science direct.com/science/article/pii/S0022103116304024.

6. Crystal D. Oberle et al., "The Galileo Bias: A Naive Conceptual Belief That Influences People's Perceptions and Performance in a Ball-Dropping Task," *Journal of Experimental Psychology, Learning, Memory, and Cognition* 31, no. 4 (2005): 643–653.

7. Brendan Nyhan et al., "Effective Messages in Vaccine Promotion: A Randomized Trial," *Pediatrics* 133, no. 4 (April 2014), http://pediatrics.aap publications.org/content/133/4/e835.long.

8. The discussion on the loss of the Mars Climate Orbiter is based on the following sources: Squyres, *Roving Mars*; Oberg, "Mars Probe Went off Course"; Euler, Jolly, and Curtis, "Failures of the Mars Climate Orbiter"; Mars Climate Orbiter Mishap Investigation Board Phase I Report, November 10, 1999, https://llis.nasa.gov/llis_lib/pdf/1009464main1_0641-mr.pdf; House Committee on Science, Space, and Technology, "Testimony of Thomas Young"; Mark Adler, interview with author, August 2018.

9. Oberg, "Mars Probe Went off Course."

10. Oberg, "Mars Probe Went off Course."

11. Richard P. Feynman, as told to Ralph Leighton and edited by Edward Hutchings, *"Surely You're Joking, Mr. Feynman!" Adventures of a Curious Character* (New York: W. W. Norton & Company, 1985), 343.

12. Sarah Scoles, *Making Contact: Jill Tarter and the Search for Extraterrestrial Intelligence* (New York: Pegasus Books, 2017).

13. John Noble Wilford, "In 'Contact,' Science and Fiction Nudge Close Together," *New York Times*, July 20, 1997, www.nytimes.com/1997/07/20/movies/in-contact-science-and-fiction-nudge-close-together.html?mtrref=www.google.com.

14. T. C. Chamberlin, "The Method of Multiple Working Hypotheses," *Science* (old series) 15, no. 92 (1890), reprinted in *Science*, May 7, 1965, available at http://arti.vub.ac.be/cursus/2005-2006/mwo/chamberlin1890science.pdf.

15. The discussion on the Mars Polar Lander is based on the following sources: NASA, "About the Deep Space Network," https://deepspace.jpl.nasa.gov/about; Dawn Levy, "Scientists Keep Searching for a Signal from Mars Polar Lander," NASA, February 1, 2000, https://mars.jpl.nasa.gov/msp98/news/mpl000201.html; Squyres, *Roving Mars*; NASA, "Listening for Mars Polar Lander," *NASA Science*, January 31, 2000, https://science.nasa.gov/science-news/science-at-nasa/2000/ast01feb_1; Natasha Mitchell, "Sweet Whispers from Mars Could Be Polar Lander," *ABC Science*, January 28, 2000, www.abc.net.au/science/articles/2000/01/28/96225.htm.

16. Levy, "Scientists Keep Searching."

17. Squyres, *Roving Mars*, 68.

18. Squyres, *Roving Mars*, 70.

19. Francis Bacon, *Novum Organum* (1902), 24.

20. Levy, "Scientists Keep Searching."

21. Kenneth L. Corum and James F. Corum, "Nikola Tesla and the Planetary Radio Signals," 2003, www.teslasociety.com/mars.pdf.

22. Chamberlin, "Multiple Working Hypotheses."

23. Robertson Davies, *Tempest-Tost* (New York: Rinehart, 1951).

24. Chamberlin, "Multiple Working Hypotheses."

25. F. Scott Fitzgerald, "The Crack-Up," *Esquire*, February, March, and April 1936 and reprinted March 7, 2017, www.esquire.com/lifestyle/a4310/the-crack-up/#ixzz1Fvs5lu8w.

26. Sarah Charley, "What's Really Happening During an LHC Collision?," *Symmetry*, June 30, 2017, www.symmetrymagazine.org/article/whats-really-happening-during-an-lhc-collision.

27. Charley, "LHC Collision?"

28. Charley, "LHC Collision?"

29. Bill Demain, "How Malfunctioning Sharks Transformed the Movie Business," *Mental Floss*, June 20, 2015, https://mentalfloss.com/article/31105/how-steven-spielbergs-malfunctioning-sharks-transformed-movie-business.

30. Robert Cialdini, *Pre-Suasion: A Revolutionary Way to Influence and Persuade* (New York: Simon & Schuster, 2016), 22.

31. Daniel Simmons and Christopher Chabris, "Selective Attention Test," video, YouTube, uploaded March 10, 2010, www.youtube.com/watch?v=vJG698U2Mvo.

32. Daniel Simmons and Christopher Chabris, "Gorilla Experiment," Invisible Gorilla website, 2010, www.theinvisiblegorilla.com/gorilla_experiment.html; Christopher Chabris and Daniel Simmons, *The Invisible Gorilla: And Other Ways Our Intuitions Deceive Us* (New York: Crown, 2010).

33. Euler, Jolly, and, Curtis, "Failures of the Mars Climate Orbiter."

34. Sir Arthur Conan Doyle, "Adventure 1: Silver Blaze," in *The Memoirs of Sherlock Holmes* (New York, 1894).

35. P. C. Wason, "On the Failure to Eliminate Hypotheses in a Conceptual Task," *Quarterly Journal of Experimental Psychology* 12, no. 3 (July 1, 1960): 129–140, https://pdfs.semanticscholar.org/86db/64c600fe59acfc48fd22bc8484485d5e7337.pdf.

36. "Peter Wason," obituary, *(London) Telegraph*, April 22, 2003, www.telegraph.co.uk/news/obituaries/1428079/Peter-Wason.html.

37. Alan Lightman, *Searching for Stars on an Island in Maine* (New York: Pantheon Books, 2018).

38. Chris Kresser, "Dr. Chris Shade on Mercury Toxicity," *Revolution Health Radio*, May 21, 2019, https://chriskresser.com/dr-chris-shade-on-mercury-toxicity.

39. Gary Taubes, "Do We Really Know What Makes Us Healthy?," *New York Times*, September 16, 2007, www.nytimes.com/2007/09/16/magazine/16epidemiology-t.html.

40. Carl Sagan, *The Demon-Haunted World: Science as a Candle in the Dark* (New York: Random House, 1995; reprint Ballantine, 1997), 211.

41. Vox, "Why Elon Musk Says We're Living in a Simulation," video, YouTube, uploaded August 15, 2016, www.youtube.com/watch?v=J0KHiiTtt4w.

42. Hal Gregersen, "Bursting the CEO Bubble," *Harvard Business Review*, April 2017, https://hbr.org/2017/03/bursting-the-ceo-bubble.

43. Shane Parrish, "How Darwin Thought: The Golden Rule of Thinking," Farnam Street (blog), January 2016, https://fs.blog/2016/01/charles-darwin-thinker.

44. Michael Lewis, "The King of Human Error," *Vanity Fair*, November 8, 2011, www.vanityfair.com/news/2011/12/michael-lewis-201112.

45. Lewis, "King of Human Error."

46. Charles Thompson, "Harlan's Great Dissent," *Kentucky Humanities* 1 (1996), https://louisville.edu/law/library/special-collections/the-john-marshall-harlan-collection/harlans-great-dissent.

47. Thompson, "Harlan's Great Dissent."

48. Walter Isaacson, *Leonardo da Vinci* (New York: Simon & Schuster, 2017), 435.

49. Gregersen, "Bursting the CEO Bubble."

50. Emmanuel Trouche et al., "The Selective Laziness of Reasoning," *Cognitive Science* 40, no. 6 (November 2016): 2122–2136, www.ncbi.nlm.nih.gov/pubmed/26452437.

51. Elizabeth Kolbert, "Why Facts Don't Change Our Minds," *New Yorker*, February 19, 2017, www.newyorker.com/magazine/2017/02/27/why-facts-dont-change-our-minds.

52. "Peter Wason," obituary.

53. James Robert Brown, *The Laboratory of the Mind: Thought Experiments in the Natural Sciences* (New York: Routledge, 1991), 20.

54. Manjit Kumar, *Quantum: Einstein, Bohr, and the Great Debate About the Nature of Reality* (New York: W.W. Norton, 2009); Carlo Rovelli, *Seven Brief Lessons on Physics*, trans. Simon Carnell and Erica Segre (New York: Riverhead Books, 2016).

55. Thomas Schelling, "The Role of War Games and Exercises," in *Managing Nuclear Operations*, ed. A. Carter, J. Steinbruner, and C. Zraket (Washington, DC: Brookings Institution, 1987), 426–444.

56. John D. Barrow, Paul C. W. Davies, and Charles L. Harper Jr., eds., *Science and Ultimate Reality: Quantum Theory, Cosmology, and Complexity* (New York: Cambridge University Press, 2004), 3.

57. David Foster Wallace, "This Is Water," commencement address at Kenyon College, Gambier, OH, May 21, 2005.

58. Errol Morris, "The Anosognosic's Dilemma: Something's Wrong but You'll Never Know What It Is," Opinionator, *New York Times*, June 24, 2010, https://opinionator.blogs.nytimes.com/2010/06/24/the-anosognosics-dilemma-somethings-wrong-but-youll-never-know-what-it-is-part-5.

59. Stanford Graduate School of Business, "Marc Andreessen on Change, Constraints, and Curiosity," video, YouTube, uploaded November 14, 2016, www.youtube.com/watch?v=P-T2VAcHRoE&feature=youtu.be.

60. Chip Heath and Dan Heath, *Decisive: How to Make Better Choices in Life and Work* (New York: Crown Business, 2013).

61. Shane Parrish, "The Work Required to Have an Opinion," Farnam Street (blog), April 2013, https://fs.blog/2013/04/the-work-required-to-have-an-opinion.

62. Rovelli, *Seven Brief Lessons on Physics*, 21.

Chapter 7: Test as You Fly, Fly as You Test

1. The opening section on healthcare.gov is based on the following sources: Sharon LaFraniere and Eric Lipton, "Officials Were Warned About Health Site Woes," *New York Times*, November 18, 2013, www.nytimes.com/2013/11/19/us/politics/administration-open-to-direct-insurance-company-signups.html; Frank Thorp, "'Stress Tests' Show Healthcare.gov Was Overloaded," NBC News, November 6, 2013, www.nbcnews.com/politics/politics-news/stress-tests-show-healthcare-gov-was-overloaded-flna8C11548230; Amy Goldstein, "HHS Failed to Heed Many Warnings That HealthCare.gov Was in Trouble," *Washington Post*, February 23, 2016, www.washingtonpost.com/national/health-science/hhs-failed-to-heed-many-warnings-that-healthcaregov-was-in-trouble/2016/02/22/dd344e7c-d67e-11e5-9823-02b905009f99_story.html?noredirect=on&utm_term=.b81dd6679eee; Wyatt Andrews and Anna Werner, "Healthcare.gov Plagued by Crashes on 1st Day," *CBS News*, October 1, 2013, www.cbsnews.com/news/healthcaregov-plagued-by-crashes-on-1st-day; Adrianne Jeffries, "Why Obama's Healthcare.gov Launch Was Doomed to Fail," *Verge*, October 8, 2013, www.theverge.com/2013/10/8/4814098/why-did-the-tech-savvy-obama-administration-launch-a-busted-healthcare-website; "The Number 6 Says It All About the HealthCare.gov Rollout," NPR, December 27, 2013, www.npr.org/sections/health-shots/2013/12/27/257398910/the-number-6-says-it-all-about-the-healthcare-gov-rollout; Kate Pickert, "Report: Cost of HealthCare.Gov Approaching \$1 Billion," *Time*, July 30, 2014, http://time.com/3060276/obamacare-affordable-care-act-cost.

2. Marshall Fisher, Ananth Raman, and Anna Sheen McClelland, "Are You Ready?," *Harvard Business Review*, August 2000) https://hbr.org/2000/07/are-you-ready.

3. Fisher, Raman, and McClelland, "Are You Ready?"

4. Richard Feynman, Messenger Lectures, Cornell University, BBC, 1964, www.cornell.edu/video/playlist/richard-feynman-messenger-lectures/player.

5. NASA Jet Propulsion Laboratory, "The FIDO Rover," NASA, https://www-robotics.jpl.nasa.gov/systems/system.cfm?System=1.

6. NASA, "Space Power Facility," www1.grc.nasa.gov/facilities/sec.

7. The discussion on the airbag tests for the Mars Exploration Rovers is based on the following sources: Steve Squyres, *Roving Mars: Spirit, Opportunity, and the Exploration of the Red Planet* (New York: Hyperion, 2005); Adam Steltzner and William Patrick, *Right Kind of Crazy: A True Story of Teamwork, Leadership, and High-Stakes Innovation* (New York: Portfolio/Penguin, 2016).

8. NASA, "Calibration Targets," https://mars.nasa.gov/mer/mission/instruments/calibration-targets.

9. "Interview with Bill Nye: The Sundial Guy," *Astrobiology Magazine*, October 8, 2003, www.astrobio.net/mars/interview-with-bill-nye-the-sundial-guy.

10. Donella Meadows, *Thinking in Systems: A Primer* (White River Junction, VT: Chelsea Green Pub., 2008), 12.

11. Kim Lane Scheppele, "The Rule of Law and the Frankenstate: Why Governance Checklists Do Not Work," *Governance: An International Journal of Policy, Administration, and Institutions* 26, no. 4 (October 2013): 559–562, https://online library.wiley.com/doi/pdf/10.1111/gove.12049.

12. Lorraine Boissoneault, "The True Story of the Reichstag Fire and the Nazi Rise to Power," *Smithsonian Magazine*, February 21, 2017, www.smithsonian mag.com/history/true-story-reichstag-fire-and-nazis-rise-power-180962240; John Mage and Michael E. Tigar, "The Reichstag Fire Trial, 1933–2008: The Production of Law and History," *Monthly Review*, March 1, 2009, http://monthly review.org/2009/03/01/the-reichstag-fire-trial-1933-2008-the-production-of-law-and-history.

13. The discussion on the design flaw of the Mars Polar Lander is based on Squyres, *Roving Mars*, 63–64.

14. US Department of Health and Human Services, Office of Inspector General, "An Overview of 60 Contracts That Contributed to the Development and Operation of the Federal Marketplace," August 2014, https://oig.hhs.gov/oei/reports/oei-03-14-00231.pdf.

15. The discussion on the tests conducted on Air Force volunteers is based on Mary Roach, *Packing for Mars: The Curious Science of Life in the Void* (New York: W.W. Norton, 2010).

16. AviationCV.com, "G-Force Process on Human Body," *Aerotime News Hub*, January 13, 2016, www.aviationcv.com/aviation-blog/2016/2721.

17. The discussion on Ham the Chimp is based on Roach, *Packing for Mars*.

18. Roach, *Packing for Mars*.

19. NASA, "Selection and Training of Astronauts," https://science.ksc.nasa .gov/mirrors/msfc/crew/training.html.

20. NASA, "Zero-Gravity Plane on Final Flight," October 29, 2004, www .nasa.gov/vision/space/preparingtravel/kc135onfinal.html.

21. NASA, "Selection and Training of Astronauts."

22. Eric Berger, "Why Is NASA Renting Out Its Huge Astronaut Pool? To Keep the Lights Turned On," *Ars Technica*, February 8, 2017, https://arstechnica .com/science/2017/02/as-it-seeks-to-pare-costs-nasa-opens-its-historic-facilities -to-private-companies.

23. Chris Hadfield, *An Astronaut's Guide to Life on Earth: What Going to Space Taught Me About Ingenuity, Determination, and Being Prepared for Anything* (New York: Little, Brown and Company, 2013).

24. Roach, *Packing for Mars*.

25. Robert Kurson, *Rocket Men: The Daring Odyssey of Apollo 8 and the Astronauts Who Made Man's First Journey to the Moon* (New York: Random House, 2018).

26. NASA, "Selection and Training of Astronauts."

27. Kurson, *Rocket Men*.

28. Craig Nelson, *Rocket Men: The Epic Story of the First Men on the Moon* (New York: Viking, 2009).

29. Hadfield, *An Astronaut's Guide*.

30. *In the Shadow of the Moon*, directed by Dave Sington (Velocity/Think Film, 2008), DVD.

31. Michael Roberto, Richard M. J. Bohmer, and Amy C. Edmondson, "Facing Ambiguous Threats," *Harvard Business Review*, November 2006, https://hbr .org/2006/11/facing-ambiguous-threats; Rebecca Wright et al., *Johnson Space Center Oral History Project* (Washington, DC: NASA, January 8, 1999), https:// history.nasa.gov/SP-4223/ch6.htm.

32. Neel V. Patel, "The Greatest Space Hack Ever," *Popular Science*, October 8, 2014, www.popsci.com/article/technology/greatest-space-hack-ever#page-2.

33. The discussion on Chief Justice John Roberts's preparation strategy is based on the following sources: Roger Parloff, "On History's Stage: Chief Justice John Roberts Jr.," *Fortune*, June 3, 2011, http://fortune.com/2011/01/03/on-historys -stage-chief-justice-john-roberts-jr; Bryan Garner, "Interviews with United States Supreme Court Justices," in *Scribes Journal of Legal Writing* (Lansing, MI, 2010), 7, https://legaltimes.typepad.com/files/garner-transcripts-1.pdf; Charles Lane, "Nominee Excelled as an Advocate Before Court," *Washington Post*, July 24, 2005, www.washingtonpost.com/wp-dyn/content/article/2005/07/23/AR2005072300 881_2.html.

34. The discussion on Amelia Boone's training is based on the following sources: Tom Bilyeu, "How to Cultivate Toughness: Amelia Boone on Impact Theory,"

video, YouTube, uploaded March 7, 2017, www.youtube.com/watch?v=_J49o G5MnN4; Marissa Stephenson, "Amelia Boone Is Stronger than Ever," *Runner's World*, June 19, 2018, www.runnersworld.com/runners-stories/a20652405/amelia -boone-is-stronger-than-ever; "Altra Signs Amelia Boone—World Champion Obstacle Course Racer and Ultrarunner," *Endurance Sportswire*, January 18, 2019, www.endurancesportswire.com/altra-signs-amelia-boone-world-champion -obstacle-course-racer-and-ultrarunner; Melanie Mitchell, "Interview with OCR World Champion Amelia Boone," *JackRabbit*, December 12, 2017, www.jack rabbit.com/info/blog/interview-with-ocr-world-champion-amelia-boone.

35. Tough Mudder, "World's Toughest Mudder," https://toughmudder.com /events/2019-worlds-toughest-mudder; Simon Donato, "Ten Tips on How to Beat the World's Toughest Mudder," *Huffington Post*, December 6, 2017, www .huffpost.com/entry/ten-tips-on-how-to-beat-t_b_8143862.

36. Roberto, Bohmer, and Edmondson, "Facing Ambiguous Threats."

37. The discussion on the iPhone is based on Derek Thompson, *Hit Makers: The Science of Popularity in an Age of Distraction* (New York: Penguin, 2018), 232–233.

38. The discussion on George Gallup is based on Thompson, *Hit Makers*.

39. Amy Kaufman, "Chris Rock Tries Out His Oscar Material at the Comedy Store," *Los Angeles Times*, February 26, 2016, www.latimes.com/entertainment /la-et-mn-chris-rock-oscars-monologue-comedy-store-20160226-story.html.

40. Jess Zafarris, "Jerry Seinfeld's 5-Step Comedy Writing Process," *Writer's Digest*, May 13, 2019, www.writersdigest.com/writing-articles/by-writing-genre /humor/jerry-seinfelds-5-step-comedy-writing-process; Daniel Auld, "What Does UX and Stand-Up Comedy Have in Common? More Than You Realize," UX Collective, August 1, 2018, https://uxdesign.cc/what-does-ux-and-stand -up-comedy-have-in-common-more-than-you-realise-d18066aeaecf.

41. Entrepreneurship.org, "Field Observations with Fresh Eyes: Tom Kelley (IDEO)," video, YouTube, uploaded June 24, 2011, www.youtube.com/watch ?v=tvkivmyKgEA.

42. Paul Bennett, "Design Is in the Details," TED talk, July 2005, www.ted .com/talks/paul_bennett_finds_design_in_the_details.

43. Art Kleiner, "The Thought Leader Interview: Tim Brown," *Strategy + Business*, August 27, 2009, www.strategy-business.com/article/09309?gko=84f90.

44. Kleiner, "Tim Brown."

45. "Ideo on *60 Minutes* and *CBS This Morning*," video, IDEO, April 2013, www.ideo.com/post/ideo-on-60-minutes-and-cbs-this-morning.

46. Joe Rogan, "Neil deGrasse Tyson," episode 919, video, Joe Rogan Experience Podcast, February 21, 2017, http://podcasts.joerogan.net/podcasts/neil -degrasse-tyson.

47. The discussion on Seinfeld is based on Thompson, *Hit Makers*.

48. The discussion on Clever Hans is based on Stuart Firestein, *Ignorance: How It Drives Science* (New York: Oxford University Press, 2012), 94–95.

49. Tim Ferriss, "Cal Fussman Corners Tim Ferriss," episode 324 (transcript), *The Tim Ferriss Show*, https://tim.blog/2018/07/05/the-tim-ferriss-show -transcripts-cal-fussman-corners-tim-ferriss; Tim Ferriss, interview with author, May 2019.

50. The discussion on the Hubble Space Telescope is based on Arthur Fisher, "The Trouble with Hubble," *Popular Science*, October 1990; Lew Allen et al., "The Hubble Space Telescope Optical Systems Failure Report," NASA, November 1990, https://ntrs.nasa.gov/archive/nasa/casi.ntrs.nasa.gov/19910003124.pdf; NASA, "About the Hubble Space Telescope," updated December 18, 2018, www .nasa.gov/mission_pages/hubble/story/index.html; Nola Taylor Redd, "Hubble Space Telescope: Pictures, Facts & History," *Space.com*, December 15, 2017, www .space.com/15892-hubble-space-telescope.html; NASA, "Hubble's Mirror Flaw," www.nasa.gov/content/hubbles-mirror-flaw.

51. Ozan Varol, "Julie Zhuo on Becoming a Facebook Manager at 25, Overcoming the Impostor Syndrome, and Staying in the Discomfort Zone," Famous Failures (podcast), March 25, 2019, https://ozanvarol.com/julie-zhuo.

Chapter 8: Nothing Succeeds Like Failure

1. Suzanne Deffree, "1st US Satellite Attempt Fails, December 6, 1957," EDN Network, December 6, 2018, www.edn.com/electronics-blogs/edn-moments /4402889/1st-US-satellite-attempt-fails--December-6--1957.

2. Richard Hollingham, "The World's Oldest Scientific Satellite Is Still in Orbit," BBC, October 6, 2017, www.bbc.com/future/story/20171005-the-worlds -oldest-scientific-satellite-is-still-in-orbit.

3. Loyd S. Swenson Jr, James M. Grimwood, and Charles C. Alexander, "Little Joe Series," in *This New Ocean: A History of Project Mercury* (Washington, DC: NASA, 1989), https://history.nasa.gov/SP-4201/ch7-7.htm.

4. NASA, "MR-1: The Four-Inch Flight," in *This New Ocean: A History of Project Mercury* (Washington, DC: NASA, 1989), https://history.nasa.gov /SP-4201/ch9-7.htm.

5. Jeffrey Kluger, "On TIME's Podcast 'Countdown:' The Flight That Nearly Took Neil Armstrong's Life," *Time*, July 31, 2017, http://time.com/4880012 /neil-armstrong-apollo-gemini-nasa.

6. FailCon, "About FailCon," http://thefailcon.com/about.html; FuckUp Nights, https://fuckupnights.com.

7. Shane Snow, *Smartcuts: The Breakthrough Power of Lateral Thinking* (New York: HarperBusiness, 2014), Kindle.

8. Gene Kranz, *Failure Is Not an Option: Mission Control From Mercury to Apollo 13 and Beyond* (New York: Simon & Schuster, 2009), 12.

9. Jennifer Reingold, "Hondas in Space," *Fast Company*, February 1, 2005, www.fastcompany.com/52065/hondas-space.

10. Chuck Salter, "Failure Doesn't Suck," *Fast Company*, May 1, 2007, www .fastcompany.com/59549/failure-doesnt-suck.

11. Hans C. Ohanian, *Einstein's Mistakes: The Human Failings of Genius* (New York: W.W. Norton & Company, 2009).

12. Jillian D'Onfro, "Jeff Bezos: Why It Won't Matter If the Fire Phone Flops," *Business Insider*, December 2, 2014, www.businessinsider.com/jeff-bezos-on-big -bets-risks-fire-phone-2014-12.

13. D'Onfro, "If the Fire Phone Flops."

14. Derek Thompson, "Google X and the Science of Radical Creativity," *Atlantic*, November 2017, www.theatlantic.com/magazine/archive/2017/11/x-google -moonshot-factory/540648.

15. Astro Teller, "The Head of 'X' Explains How to Make Audacity the Path of Least Resistance," *Wired*, April 15, 2016, www.wired.com/2016/04/the-head-of-x -explains-how-to-make-audacity-the-path-of-least-resistance.

16. Adele Peters, "Why Alphabet's Moonshot Factory Killed Off a Brilliant Carbon-Neutral Fuel," *Fast Company*, October 13, 2016, www.fastcompany .com/3064457/why-alphabets-moonshot-factory-killed-off-a-brilliant-carbon -neutral-fuel.

17. Adam Grant, *Originals: How Non-Conformists Move the World* (New York: Viking, 2017), 37.

18. Grant, *Originals*.

19. Grant, *Originals*.

20. Grant, *Originals*.

21. Emma Brockes, "Tom Hanks: 'I've Made a Lot of Movies That Didn't Make Sense—or Money,'" *Guardian*, October 14, 2017, www.theguardian.com /film/2017/oct/14/tom-hanks-movies-didnt-make-sense-or-money-interview -short-stories.

22. Paul Gompers et al., "Performance Persistence in Entrepreneurship," *Journal of Financial Economics* 96 (2010): 18–32.

23. K. C. Diwas, Bradley R. Staats, and Francesca Gino, "Learning from My Success and from Others' Failure: Evidence from Minimally Invasive Cardiac Surgery," *Management Science* 59, no. 11 (June 14, 2013): 2413–2634, https://pubs online.informs.org/doi/abs/10.1287/mnsc.2013.1720.

24. Steve Squyres, *Roving Mars: Spirit, Opportunity, and the Exploration of the Red Planet* (New York: W.W. Norton, 2005), 10.

25. University of California Television, "Roving Mars with Steve Squyres: Conversations with History," video, YouTube, uploaded August 18, 2011, www .youtube.com/watch?v=NI6KEzsb26U&feature=youtu.be; Dian Schaffhauser, "Steven Squyres Doesn't Mind Failure: An Interview with the Scientist Behind the Mars Rovers," MPUG [Microsoft Project User Group], February 9, 2016, www.mpug.com/articles/steven-squyres-interview.

26. Squyres, *Roving Mars*, 138.

27. Squyres, *Roving Mars*, 156–163.

28. Squyres, *Roving Mars*, 203–217.

29. Stephen Jay Gould, *The Panda's Thumb: More Reflections in Natural History* (New York: W. W. Norton & Company, 1980; reissued paperback, 1992), 244.

30. B. C. Forbes, "Why Do So Many Men Never Amount to Anything?," *American Magazine*, January 1921.

31. T. H. White, *The Once and Future King* (New York: Penguin Group, 2011).

32. The discussion on the Falcon 1 is based on the following sources: Tim Fernholz, *Rocket Billionaires: Elon Musk, Jeff Bezos, and the New Space Race* (Boston: Houghton Mifflin Harcourt, 2018); Snow, *Lateral Thinking*; Chris Bergin, "Falcon I Flight: Preliminary Assessment Positive for SpaceX," *Spaceflight.com*, March 24, 2007, www.nasaspaceflight.com/2007/03/falcon-i-flight-preliminary -assessment-positive-for-spacex; Tim Fernholz, "What It Took for Elon Musk's SpaceX to Disrupt Boeing, Leapfrog NASA, and Become a Serious Space Company," *Quartz*, October 21, 2014, https://qz.com/281619/what-it-took-for-elon -musks-spacex-to-disrupt-boeing-leapfrog-nasa-and-become-a-serious-space -company; Max Chafkin, "SpaceX's Secret Weapon Is Gwynne Shotwell," *Bloomberg Quint*, July 26, 2018, www.bloombergquint.com/businessweek/she -launches-spaceships-sells-rockets-and-deals-with-elon-musk; Elon Musk, "Falcon 1, Flight 3 Mission Summary," SpaceX, August 6, 2008, www.spacex.com /news/2013/02/11/falcon-1-flight-3-mission-summary; Dolly Singh, "What Is It Like to Work with Elon Musk?," *Slate*, August 14, 2013, https://slate.com/human -interest/2013/08/elon-musk-what-is-it-like-to-work-for-the-spacex-tesla-chief .html; Tom Junod, "Elon Musk: Triumph of His Will," *Esquire*, November 15, 2012, www.esquire.com/news-politics/a16681/elon-musk-interview-1212.

33. Snow, *Lateral Thinking*.

34. F. Scott Fitzgerald, *Tender Is the Night* (1934; repr., New York: Scribner's, 1977).

35. Andre Agassi, *Open: An Autobiography* (New York: Vintage Books, 2010), 372.

36. Ed Catmull, *Creativity, Inc.: Overcoming the Unseen Forces That Stand in the Way of True Inspiration* (New York: Random House, 2014).

37. Shane Parrish, "Your First Thought Is Rarely Your Best Thought: Lessons on Thinking," Farnam Street (blog), February 2018, https://fs.blog/2018/02/first -thought-not-best-thought.

38. Chris Hadfield, *An Astronaut's Guide to Life on Earth: What Going to Space Taught Me About Ingenuity, Determination, and Being Prepared for Anything* (New York: Little, Brown and Company, 2013).

39. Parrish, "Your First Thought."

40. Ben Horowitz, "Lead Bullets," Andreessen Horowitz, November 13, 2011, https://a16z.com/2011/11/13/lead-bullets.

41. Annie Duke, *Thinking in Bets: Making Smarter Decisions When You Don't Have All the Facts* (New York: Portfolio/Penguin, 2018).

42. Lars Lefgren, Brennan Platt, and Joseph Price, "Sticking with What (Barely) Worked: A Test of Outcome Bias," *Management Science* 61 (2015): 1121–1136.

43. James D. Watson, *A Passion for DNA: Genes, Genomes, and Society* (Cold Spring Harbor, NY: Cold Spring Harbor Laboratory Press, 2001), 44.

44. Jeff Dyer and Hal Gregersen, "How Does Amazon Stay at Day One?," *Forbes*, August 8, 2017, www.forbes.com/sites/innovatorsdna/2017/08/08/how-does-amazon-stay-at-day-one/#36d005d67e4d.

45. Tim Ferriss, "Maria Sharapova," episode 261 (transcript), *Tim Ferriss Show*, May 30, 2018, https://tim.blog/2018/05/30/tim-ferriss-show-transcript-maria-sharapova.

46. Elizabeth Gilbert, *Big Magic: Creative Living Beyond Fear* (New York: Riverhead Books, 2015), 259.

47. Steven Levy, "Google Glass 2.0 Is a Startling Second Act," *Wired*, July 18, 2017, www.wired.com/story/google-glass-2-is-here.

48. Heather Hargreaves, "How Google Glass Will Change How You Do Business," *Entrepreneur Handbook*, March 25, 2019.

49. Ian Osterloh, "How I Discovered Viagra," *Cosmos*, April 27, 2015, https://cosmosmagazine.com/biology/how-i-discovered-viagra; Jacque Wilson, "Viagra: The Little Blue Pill That Could," CNN, March 27, 2013, www.cnn.com/2013/03/27/health/viagra-anniversary-timeline/index.html.

50. The discussion on Mike Nichols is based on Gilbert, *Big Magic*, 246.

51. Rosamund Stone Zander and Benjamin Zander, *The Art of Possibility: Transforming Professional and Personal Life* (Boston: Harvard Business School Press, 2000), 31.

52. Union of Concerned Scientists, "Voices of Federal Scientists: Americans' Health and Safety Depends on Independent Science," January 2009, 2, www.ucsusa.org/sites/default/files/legacy/assets/documents/scientific_integrity/Voices_of_Federal_Scientists.pdf.

53. Jennifer J. Kish-Gephart et al., "Silenced by Fear," *Research in Organizational Behavior* 29 (December 2009): 163–193, www.researchgate.net/publication/238382691_Silenced_by_fear.

54. NASA, "Mars Polar Lander Fact Sheet," https://mars.nasa.gov/msp98/lander/fact.html.

55. Hadfield, *Astronaut's Guide*, 81–83.

56. Diwas, Staats, and Gino, "Learning from My Success."

57. Ed Catmull and Amy Wallace, *Creativity, Inc.: Overcoming the Unseen Forces That Stand in the Way of True Inspiration* (Toronto: Random House Canada, 2014), 123.

58. David W. Bates et al., "Relationship Between Medication Errors and Adverse Drug Events," *Journal of General Internal Medicine* 10, no. 4 (April 1995): 199–205, www.ncbi.nlm.nih.gov/pubmed/7790981.

59. Amy C. Edmondson, "Learning from Mistakes Is Easier Said than Done: Group and Organizational Influences on the Detection and Correction of Human Error," *Journal of Applied Behavioral Science* 32, no. 1 (1996): 5–28.

60. Amy C. Edmondson, "Managing the Risk of Learning: Psychological Safety in Work Teams," in *International Handbook of Organizational Teamwork and Cooperative Learning*, ed. Michael A. West, Dean Tjosvold, and Ken G. Smith (West Sussex, UK: John Wiley & Sons, 2003).

61. Neil Robert Anderson, "Innovation in Top Management Teams," *Journal of Applied Psychology* 81, no. 6 (December 1996): 680–693; Amy C. Edmondson, Richard Bohmer, and Gary Pisano, "Learning New Technical and Interpersonal Routines in Operating Room Teams," in *Research on Managing Groups and Teams: Technology*, ed. B. Mannix, M. Neale, and T. Griffith (Stamford, CT: JAI Press, 2000) 3: 29–51; Amy C. Edmondson, Richard Bohmer, and Gary Pisano, "Disrupted Routines: Team Learning and New Technology Implementation in Hospitals," *Administrative Science Quarterly* 46 (December 2001): 685–716; Charlene D'Andrea-O'Brien and Anthony Buono, "Building Effective Learning Teams: Lessons from the Field," *Society for the Advancement of Management Journal* 61, no. 3 (1996).

62. Amy C. Edmondson, "Psychological Safety and Learning Behavior in Work Teams," *Administrative Science Quarterly* 44, no. 2 (June 1999): 350–383.

63. Edmondson, Bohmer, and Pisano, "Interpersonal Routines in Operating Room Teams."

64. Edmondson, "Learning from Mistakes."

65. Derek Thompson, "Google X and the Science of Radical Creativity," *Atlantic*, November 2017, www.theatlantic.com/magazine/archive/2017/11/x-google-moonshot-factory/540648.

66. Astro Teller, "The Head of 'X' Explains How to Make Audacity the Path of Least Resistance," *Wired*, April 15, 2016, www.wired.com/2016/04/the-head-of-x-explains-how-to-make-audacity-the-path-of-least-resistance/#.2vy7nkes6.

67. Obi Felten, "How to Kill Good Things to Make Room for Truly Great Ones," X Blog, March 8, 2016, https://blog.x.company/how-to-kill-good-things-to-make-room-for-truly-great-ones-867fb6ef026.

68. Dyer and Gregersen, "How Does Amazon Stay at Day One?"

69. Tom Peters, *The Circle of Innovation: You Can't Shrink Your Way to Greatness* (New York: Vintage Books, 1999), viii.

70. Hadfield, *An Astronaut's Guide*, 79–80.

71. Mario Livio, *Brilliant Blunders: From Darwin to Einstein—Colossal Mistakes by Great Scientists That Changed Our Understanding of Life and the Universe* (New York: Simon & Schuster, 2013), 266.

72. Hal Gregersen, "Bursting the CEO Bubble," *Harvard Business Review*, April 2017, https://hbr.org/2017/03/bursting-the-ceo-bubble.

73. Catmull and Wallace, *Creativity, Inc.*

74. Tyler Cowen, "My Biggest Regret," *Econ Journal Watch*, May 2017, https://pingpdf.com/pdf-econ-journal-watch-142-may-2017.html.

75. Anna Bruk, Sabine G. Scholl, and Herbert Bless, "Beautiful Mess Effect: Self–Other Differences in Evaluation of Showing Vulnerability," *Journal of Personality and Social Psychology* 115, no. 2 (2018): 192–205, https://psycnet.apa.org/record/2018-34832-002.

76. Elliot Aronson, Ben Willerman, and Joanne Floyd, "The Effect of a Pratfall on Increasing Interpersonal Attractiveness," *Psychonomic Science* 4, no. 6 (June 1966): 227–228, https://link.springer.com/article/10.3758/BF03342263; Emily Esfahani Smith, "Your Flaws Are Probably More Attractive than You Think They Are," *Atlantic*, January 9, 2019, www.theatlantic.com/health/archive/2019/01/beautiful-mess-vulnerability/579892.

77. Tom R. Tyler and E. Allan Lind, "A Relational Model of Authority in Groups," *Advances in Experimental Social Psychology* 25 (1992): 115–191.

78. Edmondson, Bohmer, and Pisano, "Disrupted Routines."

79. Edmondson, Bohmer, and Pisano, "Disrupted Routines."

80. Edmondson, Bohner, Pisano, "Speeding Up Team Learning."

81. Edmondson, Bohner, Pisano, "Speeding Up Team Learning."

82. Lisa Bodell, *Kill the Company: End the Status Quo, Start an Innovation Revolution* (Brookline, MA: Bibliomotion, 2016), 130.

83. Jessica Bennett, "On Campus, Failure Is on the Syllabus," *New York Times*, June 24, 2017, www.nytimes.com/2017/06/24/fashion/fear-of-failure.html.

Chapter 9: Nothing Fails Like Success

1. The opening discussion on the *Challenger* disaster is based on the following sources: Trudy E. Bell and Karl Esch, "The Fatal Flaw in Flight 51-L," *IEEE Spectrum*, February 1987, https://ieeexplore.ieee.org/document/6448023; Doug G. Ware, "Engineer Who Warned of 1986 Challenger Disaster Still Racked with Guilt, Three Decades On," UPI, January 28, 2016, www.upi.com/Top_News/US/2016/01/28/Engineer-who-warned-of-1986-Challenger-disaster-still-racked-with-guilt-three-decades-on/4891454032643; Douglas Martin, "Roger Boisjoly, 73, Dies; Warned of Shuttle Danger," *New York Times*, February 3, 2012, www.nytimes.com/2012/02/04/us/roger-boisjoly-73-dies-warned-of-shuttle-danger.html; Shaun Usher, "The Result Would Be a Catastrophe," *Letters of Note*, October 27, 2009, www.lettersofnote.com/2009/10/result-would-be-catastrophe.html; Andy Cox, "Weather's Role in the Challenger Accident," Weather Channel, January 28, 2015, https://weather.com/science/space/news/space-shuttle-challenger-weather-role; Chris Bergin, "Remembering the Mis-

takes of Challenger," NASA, January 28, 2007, www.nasaspaceflight.com /2007/01/remembering-the-mistakes-of-challenger.

2. William H. Starbuck and Frances J. Milliken, "Challenger: Fine-Tuning the Odds Until Something Breaks," *Journal of Management Studies* 25, no. 4 (1988): 319–340, https://papers.ssrn.com/sol3/papers.cfm?abstract_id=2708154.

3. James Gleick, "NASA's Russian Roulette," *Baltimore Sun*, December 15, 1993, www.baltimoresun.com/news/bs-xpm-1993-12-15-1993349207-story.html.

4. The discussion on the *Columbia* disaster is based on the following sources: Michael Roberto et al., "Columbia's Final Mission," Harvard Business School Case Collection, March 2005, www.hbs.edu/faculty/Pages/item.aspx?num=32162; Tim Fernholz, *Rocket Billionaires: Elon Musk, Jeff Bezos, and the New Space Race* (Boston: Houghton Mifflin Harcourt, 2018); Elizabeth Howell, "Columbia Disaster: What Happened, What NASA Learned," *Space.com*, February 1, 2019, www .space.com/19436-columbia-disaster.html; Robert Lee Hotz, "Decoding Columbia: A Detective Story," *Los Angeles Times*, December 21, 2003, www.latimes.com /nation/la-sci-shuttle21dec21-story.html; Anna Haislip, "Failure Leads to Success," NASA, February 21, 2007, www.nasa.gov/offices/nesc/press/070221.html.

5. Fernholz, *Rocket Billionaires*, 73.

6. Amy C. Edmondson et al., "The Recovery Window: Organizational Learning Following Ambiguous Threats," in *Organization at the Limit: Lessons from the Columbia Disaster*, ed. William H. Starbuck and Moshe Farjoun (Malden, MA: Blackwell Pub., 2009).

7. Roberto et al., "Columbia's Final Mission."

8. Roberto et al., "Columbia's Final Mission."

9. Roberto et al., "Columbia's Final Mission."

10. Roberto et al., "Columbia's Final Mission."

11. Roberto et al., "Columbia's Final Mission."

12. George Bernard Shaw, *The Doctor's Dilemma* (New York: Brentano's, 1911).

13. Bill Gates, with Nathan Myhrvold and Peter Rinearson, *The Road Ahead* (New York: Penguin Books, 1995).

14. Daniel Kahneman and Dan Lovallo, "Timid Choices and Bold Forecasts: A Cognitive Perspective on Risk Taking," *Management Science* 39, no. 1 (January 1993): 17–31, http://bear.warrington.ufl.edu/brenner/mar7588/Papers/kahneman -lovallo-mansci1993.pdf.

15. Gilles Hilary and Lior Menzly, "Does Past Success Lead Analysts to Become Overconfident?," *Management Science* 52, no. 4 (April 2006): 489–500.

16. Cyril Connolly, *Enemies of Promise* (Boston, Little, Brown and Company, 1938).

17. Boyce Rensberger and Kathy Sawyer, "Challenger Disaster Blamed on O-Rings, Pressure to Launch," *Washington Post*, June 10, 1986, www.washington post.com/archive/politics/1986/06/10/challenger-disaster-blamed-on-o -rings-pressure-to-launch/6b331ca1-f544-4147-8e4e-941b7a7e47ae.

18. E. B. White, *One Man's Meat* (New York and London, Harper & Brothers, 1942), 273.

19. William H. Starbuck and Frances J. Milliken, "Challenger: Changing the Odds Until Something Breaks," in *Organizational Realities: Studies of Strategizing and Organizing*, ed. William H. Starbuck and Moshe Farjoun (Malden, MA: Blackwell Pub., 2009).

20. NASA, "President Nixon's 1972 Announcement on the Space Shuttle," https://history.nasa.gov/stsnixon.htm (emphasis added).

21. Steven J. Dick II, "Historical Background: What Were the Shuttle's Goals and Possible Configurations?," NASA, April 5, 2001, https://history.nasa.gov/sts1/pages/scota.html.

22. Michael Roberto, Richard M. J. Bohmer, and Amy C. Edmondson, "Facing Ambiguous Threats," *Harvard Business Review*, November 2006, https://hbr.org/2006/11/facing-ambiguous-threats.

23. Starbuck and Milliken, "Challenger: Changing the Odds."

24. Roberto et al., "Columbia's Final Mission."

25. Information on cutbacks is from Starbuck and Milliken, "Challenger: Changing the Odds."

26. Diane Vaughan, testimony in "Columbia Accident Investigation Board Public Hearing," Houston, April 23, 2003, http://govinfo.library.unt.edu/caib/news/report/pdf/vol6/part08.pdf.

27. Vaughan, testimony.

28. Starbuck and Milliken, "Challenger: Changing the Odds."

29. Ronald W. Reagan, "Explosion of the Space Shuttle Challenger Address to the Nation, January 28, 1986," NASA, https://history.nasa.gov/reagan12886.html.

30. Daniel Gilbert, *Stumbling on Happiness* (New York: A. A. Knopf, 2006).

31. Tom Fordyce, "How Greene Nearly Walked Away," *BBC Sport*, July 29, 2004, http://news.bbc.co.uk/sport2/hi/athletics/3934337.stm.

32. Ryan Holiday, *Ego Is the Enemy* (New York: Portfolio, Penguin, 2016) (emphasis in original).

33. Holiday, *Ego Is the Enemy*.

34. Mia Hamm with Aaron Heifetz, *Go for the Goal: A Champion's Guide to Winning in Soccer and Life* (New York: Harper, 1999).

35. Whitney Tilson, "Warren Buffett's New Words of Wisdom," *Daily Beast*, May 3, 2009, www.thedailybeast.com/warren-buffetts-new-words-of-wisdom.

36. Daniel Pink, *When: The Scientific Secrets of Perfect Timing* (New York: Riverhead Books, 2018).

37. Jonah Berger and Devin Pope, "Can Losing Lead to Winning?," *Management Science* 57, no. 5 (May 2011), https://pubsonline.informs.org/doi/abs/10.1287/mnsc.1110.1328.

38. Berger and Pope, "Can Losing Lead to Winning?"

39. Tanya Sweeney, "Happy 60th Birthday to Madonna, the Queen of Reinvention: How She Continues to Pave the Way for Women Everywhere," *Independent*, August 12, 2018, www.independent.ie/entertainment/music/happy-60th-birthday-to-madonna-the-queen-of-reinvention-how-she-continues-to-pave-the-way-for-women-everywhere-37201633.html.

40. The Netflix discussion is based on the following sources: Scott D. Anthony and Evan I. Schwartz, "What the Best Transformational Leaders Do," *Harvard Business Review*, May 8, 2017, https://hbr.org/2017/05/what-the-best-transformational-leaders-do; Bill Taylor, "How Coca-Cola, Netflix, and Amazon Learn from Failure," *Harvard Business Review*, November 10, 2017, https://hbr.org/2017/11/how-coca-cola-netflix-and-amazon-learn-from-failure.

41. Reed Hastings, "Reed Hastings: Here's Why We're Splitting Netflix in Two and Calling the DVD Business 'Qwikster,'" *Business Insider*, September 19, 2011.

42. Bill Taylor, "Coca-Cola, Netflix, and Amazon."

43. Sim B. Sitkin, "Learning Through Failure: The Strategy of Small Losses," *Research in Organizational Behavior* 14 (1992): 231–266.

44. Sim B. Sitkin and Amy L. Pablo, "Reconceptualizing the Determinants of Risk Behavior," *Academy of Management Review* 17, no. 1 (1992).

45. Jeff Stone, "Elon Musk: SpaceX 'Complacency' Contributed to Falcon 9 Crash, Falcon Heavy Rocket Debuts in 2016," *International Business Times*, January 21, 2015, www.ibtimes.com/elon-musk-spacex-complacency-contributed-falcon-9-crash-falcon-heavy-rocket-debuts-2017809.

46. Steve Forbes, tweet on Twitter, January 2, 2015, https://twitter.com/steveforbesceo/status/551091006805118977?lang=en.

47. Robin L. Dillon and Catherine H. Tinsley, "How Near-Misses Influence Decision Making Under Risk: A Missed Opportunity for Learning," *Management Science* 54, no. 8 (2008), https://pubsonline.informs.org/doi/abs/10.1287/mnsc.1080.0869.

48. Dillon and Tinsley, "Near-Misses."

49. Dillon and Tinsley, "Near-Misses."

50. Dillon and Tinsley, "Near-Misses."

51. Diane Vaughan, *The Challenger Launch Decision: Risky Technology, Culture, and Deviance at NASA* (Chicago: University of Chicago Press, 1996), 410.

52. Roberto, Bohmer, and Edmondson, "Facing Ambiguous Threats."

53. Peter M. Madsen and Vinit Desai, "Failing to Learn? The Effects of Failure and Success on Organizational Learning in the Global Orbital Launch Vehicle Industry," *Academy of Management Journal* 53, no. 3 (November 30, 2017), https://journals.aom.org/doi/10.5465/amj.2010.51467631.

54. Mark D. Cannon and Amy C. Edmondson, "Failing to Learn and Learning to Fail (Intelligently): How Great Organizations Put Failure to Work to Innovate and Improve," *Long Range Planning* 38, no. 3 (March 2004): 299–319.

55. The discussion on Tom Brady and the New England Patriots is based on Holiday, *Ego Is the Enemy*.

56. Cork Gaines, "How the Patriots Pulled Off the Biggest Steal in NFL Draft History and Landed Future Hall of Famer Tom Brady," *Business Insider*, September 10, 2015, www.businessinsider.com/patriots-tom-brady-draft-steal-2015-1.

57. Josh St. Clair, "Why Tom Brady Is So Good, According to Former NFL Quarterbacks," *Men's Health*, January 30, 2019, www.menshealth.com /entertainment/a26078069/tom-brady-super-bowl-2019-talent.

58. Derek Thompson, "Google X and the Science of Radical Creativity," *Atlantic*, November 2017, www.theatlantic.com/magazine/archive/2017/11/x-google -moonshot-factory/540648.

59. Jack Brittain and Sim B. Sitkin, "Facts, Figures, and Organizational Decisions: Carter Racing and Quantitative Analysis in the Organizational Behavior Classroom," *Journal of Management Education* 14, no. 1 (1990): 62–81, https:// journals.sagepub.com/doi/abs/10.1177/105256298901400108.

60. "Simply Great: Charlie Munger's Speech to the Harvard School, June 1986—'Invert, Always Invert,'" *BizNews*, June 13, 1986, www.biznews.com /thought-leaders/1986/06/13/charlie-mungers-speech-to-the-harvard-school -june-1986.

61. Gary Klein, "Performing a Project Premortem," *Harvard Business Review*, September 2007, https://hbr.org/2007/09/performing-a-project-premortem.

62. Adam Smith, *The Theory of Moral Sentiments* (London: A. Millar, 1759).

63. Deborah J. Mitchell, J. Edward Russo, and Nancy Pennington, "Back to the Future: Temporal Perspective in the Explanation of Events," *Journal of Behavioral Decision Making* 2, no. 1 (January–March 1989): 25–38, https://online library.wiley.com/doi/abs/10.1002/bdm.3960020103.

64. Annie Duke, *Thinking in Bets: Making Smarter Decisions When You Don't Have All the Facts* (New York: Portfolio/Penguin, 2018).

65. "Elon Musk Answers Your Questions! SXSW, March 11, 2018," video, YouTube, uploaded March 11, 2018, www.youtube.com/watch?v=OoQARBY bkck.

66. Astro Teller, "The Head of 'X' Explains How to Make Audacity the Path of Least Resistance," *Wired*, April 15, 2016, www.wired.com/2016/04/the-head -of-x-explains-how-to-make-audacity-the-path-of-least-resistance.

67. Scott Snook and Jeffrey C. Connor, "The Price of Progress: Structurally Induced Inaction," in *Organization at the Limit: Lessons from the Columbia Disaster*, ed. William H. Starbuck and Moshe Farjoun (Malden, MA: Blackwell Pub., 2009).

68. Roger M. Boisjoly, "Ethical Decisions—Morton Thiokol and the Space Shuttle Challenger Disaster," May 15, 2006, www.onlineethics.org/Resources /thiokolshuttle/shuttle_post.aspx#publicationContent.

69. Douglas Martin, "Roger Boisjoly, 73, Dies; Warned of Shuttle Danger," *New York Times*, February 3, 2012, www.nytimes.com/2012/02/04/us/roger-boisjoly-73-dies-warned-of-shuttle-danger.html.

70. Charlan Jeanne Nemeth, "Differential Contributions of Majority and Minority Influence," *Psychological Review* 93, no. 1 (January 1986): 23–32, www.researchgate.net/publication/232513627_The_Differential_Contributions_of_Majority_and_Minority_Influence.

71. Vaughan, testimony.

72. Roberto, Bohmer, and Edmondson, "Facing Ambiguous Threats."

73. Vaughan, *The Challenger Launch Decision*, 386.

74. George Santayana, *The Life of Reason: Reason in Common Sense* (New York, C. Scribner's Sons, 1905).

75. Gerald J. S. Wilde, "Risk Homeostasis: A Theory About Risk Taking Behaviour," http://riskhomeostasis.org/home; Malcolm Gladwell, "Blowup," *New Yorker*, January 14, 1996.

76. M. Aschenbrenner and B. Biehl, "Improved Safety Through Improved Technical Measures? Empirical Studies Regarding Risk Compensation Processes in Relation to Anti-Lock Braking Systems," in *Challenges to Accident Prevention: The Issue of Risk Compensation Behavior*, Rüdiger M. Trimpop and Gerald J. S. Wilde (Groningen, Netherlands: STYX, 1994), https://trid.trb.org/view/457353.

77. Gerald J. S. Wilde, *Target Risk 3: Risk Homeostasis in Everyday Life* (2014), available at http://riskhomeostasis.org, 93–94.

78. Starbuck and Milliken, "Challenger: Changing the Odds."

79. Starbuck and Milliken, "Challenger: Changing the Odds."

Epilogue: The New World

1. Ross Anderson, "Exodus," *Aeon*, September 30, 2014, https://aeon.co/essays/elon-musk-puts-his-case-for-a-multi-planet-civilisation.

2. Paul Harris, "Neil Armstrong's Death Prompts Yearning for America's Past Glories," *Guardian*, August 27, 2012, www.theguardian.com/science/2012/aug/26/neil-armstrong-passing-us-yearning-glory.

3. Marina Koren, "What's So Special About the Next SpaceX Launch," *Atlantic*, March 1, 2019, www.theatlantic.com/science/archive/2019/03/nasa-prepares-pivotal-spacex-launch-iss/583906; Brad Tuttle, "Here's How Much It Costs for Elon Musk to Launch a SpaceX Rocket," *Money.com*, February 6, 2018, http://money.com/money/5135565/elon-musk-falcon-heavy-rocket-launch-cost.

4. Maria Stromova, "Trampoline to Space? Russian Official Tells NASA to Take a Flying Leap," ABC News, April 29, 2014, www.nbcnews.com/storyline/ukraine-crisis/trampoline-space-russian-official-tells-nasa-take-flying-leap-n92616.

5. Eric Berger, "Adrift: As NASA Seeks Next Mission, Russia Holds the Trump Card," *Houston Chronicle*, 2014, www.houstonchronicle.com/nasa/adrift/1.

6. Reuters, "NASA Puts Shuttle Launch Pad in Florida Up for Lease," May 23, 2013, www.reuters.com/article/us-usa-space-launchpad/nasa-puts -shuttle-launch-pad-in-florida-up-for-lease-idUSBRE94M16520130523?feed Type=RSS.

7. Jacey Fortin and Karen Zraick, "First All-Female Spacewalk Canceled Because NASA Doesn't Have Two Suits That Fit," *New York Times*, March 25, 2019, www.nytimes.com/2019/03/25/science/female-spacewalk-canceled.html.

8. For an excellent book that tells this story, see Julian Guthrie, *How to Make a Spaceship: A Band of Renegades, an Epic Race, and the Birth of Private Spaceflight* (New York: Penguin 2016).

9. "SpaceX Signs 20-Year Lease for Historic Launch Pad 39A," *NBC News*, April 15, 2014, www.nbcnews.com/science/space/spacex-signs-20-year-lease -historic-launch-pad-39a-n81226.

10. Amy Thompson, "NASA's Supersize Space Launch System Might Be Doomed," *Wired*, March 14, 2019, www.wired.com/story/nasas-super-sized -space-launch-system-might-be-doomed.

11. Jeff Bezos, letter to Amazon Shareholders, 2016 Ex-99.1, SEC.gov, www .sec.gov/Archives/edgar/data/1018724/000119312517120198/d373368dex991.htm (italics in original).

12. Walt Whitman, *Song of the Open Road* (New York: Limited Editions Club, 1990).

INDEX

Ozan Varol is a rocket scientist turned award-winning professor, author, and podcast host. A native of Istanbul, he moved to America to major in astrophysics at Cornell University, then served on the operations team for the 2003 Mars Exploration Rovers project. Varol later became a law professor at Lewis & Clark College and wrote *The Democratic Coup d'État*, published by Oxford University Press. Varol's articles have appeared in outlets such as the *Wall Street Journal*, *Newsweek*, *BBC*, *Time*, *CNN*, the *Washington Post*, *Slate*, and *Foreign Policy*. He blogs weekly on his website, ozanvarol.com. A sought-after public speaker, Varol has given countless radio and television interviews and has delivered keynote speeches to both small and large groups at major corporations, nonprofits, and government institutions.

PublicAffairs is a publishing house founded in 1997. It is a tribute to the standards, values, and flair of three persons who have served as mentors to countless reporters, writers, editors, and book people of all kinds, including me.

I. F. STONE, proprietor of *I. F. Stone's Weekly*, combined a commitment to the First Amendment with entrepreneurial zeal and reporting skill and became one of the great independent journalists in American history. At the age of eighty, Izzy published *The Trial of Socrates*, which was a national bestseller. He wrote the book after he taught himself ancient Greek.

BENJAMIN C. BRADLEE was for nearly thirty years the charismatic editorial leader of *The Washington Post*. It was Ben who gave the *Post* the range and courage to pursue such historic issues as Watergate. He supported his reporters with a tenacity that made them fearless and it is no accident that so many became authors of influential, best-selling books.

ROBERT L. BERNSTEIN, the chief executive of Random House for more than a quarter century, guided one of the nation's premier publishing houses. Bob was personally responsible for many books of political dissent and argument that challenged tyranny around the globe. He is also the founder and longtime chair of Human Rights Watch, one of the most respected human rights organizations in the world.

•   •   •

For fifty years, the banner of Public Affairs Press was carried by its owner Morris B. Schnapper, who published Gandhi, Nasser, Toynbee, Truman, and about 1,500 other authors. In 1983, Schnapper was described by *The Washington Post* as "a redoubtable gadfly." His legacy will endure in the books to come.

Peter Osnos, *Founder*